T0231104

Effective Opportunity Management for Projects

CENTER FOR BUSINESS PRACTICES

Editor

James S. Pennypacker

Director
Center for Business Practices
West Chester, Pennsylvania

The Superior Project Organization: Global Competency Standards and Best Practices, Frank Toney

The Superior Project Manager: Global Competency Standards and Best Practices, Frank Toney

Effective Opportunity Management for Projects: Exploiting Positive Risk, David Hillson

PM Practices

The Strategic Project Office: A Guide to Improving Organizational Performance, J. Kent Crawford

Project Management Maturity Model: Providing a Proven Path to Project Management Excellence, J. Kent Crawford

Managing Multiple Projects: Planning, Scheduling, and Allocating Resources for Competitive Advantage, James S. Pennypacker and Lowell D. Dye

ADDITIONAL VOLUMES IN PREPARATION

Effective Opportunity Management for Projects

Exploiting Positive Risk

David Hillson

Risk Doctor & Partners
Petersfield, Hampshire, United Kingdom

Taylor & Francis
Taylor & Francis Group
Boca Raton London New York

A CRC title, part of the Taylor & Francis imprint, a member of the
Taylor & Francis Group, the academic division of T&F Informa plc.

Published in 2004 by
CRC Press
Taylor & Francis Group
6000 Broken Sound Parkway NW, Suite 300
Boca Raton, FL 33487-2742

© 2004 by Taylor & Francis Group, LLC
CRC Press is an imprint of Taylor & Francis Group

No claim to original U.S. Government works
Printed in the United States of America on acid-free paper
10 9 8 7 6 5 4 3

International Standard Book Number-10: 0-8247-4808-5 (Hardcover)
International Standard Book Number-13: 978-0-8247-4808-1 (Hardcover)

This book contains information obtained from authentic and highly regarded sources. Reprinted material is quoted with permission, and sources are indicated. A wide variety of references are listed. Reasonable efforts have been made to publish reliable data and information, but the author and the publisher cannot assume responsibility for the validity of all materials or for the consequences of their use.

No part of this book may be reprinted, reproduced, transmitted, or utilized in any form by any electronic, mechanical, or other means, now known or hereafter invented, including photocopying, microfilming, and recording, or in any information storage or retrieval system, without written permission from the publishers.

Trademark Notice: Product or corporate names may be trademarks or registered trademarks, and are used only for identification and explanation without intent to infringe.

Library of Congress Cataloging-in-Publication Data

Catalog record is available from the Library of Congress

Taylor & Francis Group
is the Academic Division of Informa plc.

Visit the Taylor & Francis Web site at
http://www.taylorandfrancis.com

and the CRC Press Web site at
http://www.crcpress.com

Foreword

David Hulett

In this major book David Hillson has given us a full treatment of risk management with a special and convincing rationale for including opportunities as well as threats in the process. He has provided an interpretation of opportunity management as it should be practiced that will benefit all mature project managers. In doing this, he treats project risk management as a profession. This should not be surprising to those of us who share his view, but it may be to those who believe that of all the project management disciplines, risk management is the one that may be "optional."

There is a lively debate among risk management professionals over whether the definition of "risk" includes both threats and opportunities or is limited only to threats to objectives. Two reasons are commonly cited for the "risk-as-threat" view. First, dictionary definitions emphasize the probability that bad things might happen. Second, many organizations "optimize" their project plans to incorporate most of the opportunities that could happen, resulting in a baseline plan that can succeed only if "everything goes according to plan." No wonder then that many people exclusively identify risk with all the bad things that can happen to the perfect project plan, but not with opportunities. David Hillson helps us to see the other side and to balance our efforts to include identifying and capturing opportunities.

One of the first benefits of this book is the comprehensive discussion of the risk definition issue. Defining risk as "any uncertainty which, if it occurs, would affect one or more objectives," Dr Hillson demands that we explore and respond when we find opportunities as well as threats to the basic project plan. His review of the definitions of risk published by official bodies seems to favor this broader definition.

The second major insight of the book is that the most effective approach to project risk management is to deal with opportunities in the same process as threats. Most project risk management practitioners experience a bias on their own part or those of their sponsors toward looking at threats first, maybe exclusively. Some say, "We'll go back and look for opportunities later," but they hardly ever do. Using the same processes for managing opportunities and threats alike will be alien to some, but should be welcome to others who seek ways to make managing opportunities more accessible.

David Hillson puts the organization squarely in the middle of the process and emphasizes the need for organizational understanding and commitment to project risk management. Including opportunities should make some organizations more willing to explore proactively the uncertainties that affect their plans.

Practical issues may confound the practice of risk management for many people. Simply understanding "ordinary" risk management is hard enough. Incorporating opportunities may be a real stretch for some people and organizations. David's challenge in this book has been to put opportunities into the mix while telling us about good risk management processes generally, providing both the novice practitioner with the basic tools and the experienced professional with the added dimension of opportunity. His discussions of qualitative risk assessment and of risk response planning in particular include some of the most novel and useful concepts.

Clearly, an opportunity that is both highly likely to occur and provides the greatest benefit to the project if it were to occur should attract the greatest interest. The benefits would be more easily secured for the project by securing these "low-hanging fruit" risks, and organizations should exercise the most effort to get those. The profession has had some difficulty illustrating this concept, and David's "mirror probability and impact matrix" is an appealing graphical way to make opportunity-seeking choices evident.

Several ideas are introduced in Chapter 5 that require some added scrutiny by the profession. One of these is the notion that qualitative risk assessment can provide an assessment of total project risk "in absolute terms against some risk threshold that describes the amount of risk that stakeholders deem acceptable." It is still not entirely clear how an assessment that considers risks one at a time can produce an overall risk measure for the project. Another issue in the same chapter is the suggestion that while qualitative risk assessment is required for the majority of the projects, quantitative risk analysis is an optional extra. This may arise from the fact that quantitative risk analysis uses specialized software and statistical terms that some find hard to understand, but this need not necessarily imply that Monte Carlo simulation is too sophisticated for most organizations. Finally, Dr Hillson offers different ways to describe probability levels (e.g. 5% is either "very low" or "1:20" or "1–10%"), and a useful addition may be to consider describing in words the relationship between a project's overall condition and the associated probability of a risk occurring. For example, if a technology has not been tested at the prototype level, the probability of related risks may be assessed as 50–75%.

Another of David's main ideas is that data quality is important—all the tools he describes are simple and their use should be transparent. An organization should be willing to spend the time and resources required to gather credible data by means that can be defended. A good example of this is the Risk Breakdown Structure, one of his recent breakthrough concepts which has been widely accepted as an aid to identifying risks.

David Hillson makes some of the most important contributions in this book in risk response planning. His opportunity-centric responses (comparable to typical responses that address threats) should be incorporated in any serious risk response plan. He identifies Exploit, Share, and Enhance as the opportunity analogues to Avoid, Transfer and Mitigate. These new response strategies will help practitioners understand how to address opportunities proactively, and I believe that the profession should coalesce around these or similar definitions and develop practical ways of implementing them.

The chapter on implementation issues identifies a number of Critical Success Factors, particularly focusing on individuals' risk attitudes and corporate cultures as they are and as they may be modified to make risk management (not just opportunity management) successful. These cultural issues as well as some more practical ones are brought together in the outline of a Risk Management Maturity Model that should help organizations understand where they are and what they should do to improve.

The final chapter emphasizes ways that risk management including opportunities can be done with minimum pain and maximum benefits to the organization. For some time, risk management professionals have needed a concept of return on investment (ROI) for risk management, and this deserves further discussion, since it will be a major selling point to reluctant organizations (assuming it comes out positive).

While reading this book I have made many positive notes in the margins. The serious contributions made by David Hillson to advancing the concepts of risk management will now be available to all who read this very useful book.

David Hulett, Ph.D.
Principal, Hulett & Associates LLC
Los Angeles, California, U.S.A.
and Technical Director, PMI Risk Management SIG

Foreword

Chris Chapman

I think you should read this book. Stating the obvious, I would not have agreed to write a foreword for it if I thought otherwise. Reading this book will be good for you. But like taking risk when seeking opportunities more generally, reading this book will be good for you provided you understand that it comes with health warnings. David Hillson understands this. He asked three people each to write a foreword with this in mind, because *all four* of us have different views about what the health warnings should be, as well as different views about what are the best bits. I have one central health warning, and several other health warnings, which need to be explicit. But let us begin with why this book will be good for you.

Opportunity management is a key issue. There is currently no widely accepted approach for dealing with project opportunities, this book helps to fill the gap, and it captures part of the current debate on the future direction of project risk management. It is required reading to understand some of the directions project risk management is taking, how to keep abreast of current best practice, and some of the places where current best practice needs to get better.

This book provides a very rich understanding of alternative definitions of risk, institutional standards, and guidelines. It is a "must read" for this reason alone.

It also promotes a formal iterative approach to project risk management embracing both opportunities and threats. The iterative nature of risk management processes matters. Approaches which are not iterative are shallow or ineffective or inefficient or all three, because when the process starts it will usually not be clear where effort is most effectively applied. The project risk management processes I developed with BP International in the 1970s targeted completion of the first complete pass in 20% of the time available, leaving 80% for further complete or partial iterations. This version of the 80:20 rule worked well for BP for the decade I was involved in their North Sea and other operations worldwide. It has worked well for a wide range of organizations since. Many other authors have independently observed the importance of iterative approaches, and iterative processes are established best practice in a minority of organizations which define the leading edge in my view. But some widely recommended approaches and many operational corporate processes still fail to apply this basic idea. So I warmly endorse its promotion in this book.

David also provides useful discussions of a number of important issues, like the need to define risk in relation to one or more objectives; integrate opportunity and threat management; use "onion ring diagrams" (Figure 6–9) to understand sensitivity and what is driving risk, a prerequisite to responding effectively and efficiently; resolve strategic issues before tactical issues; and recognize the importance of response management in general.

Further, useful discussion of a range of topics is provided, including: documenting requirements; gaining stakeholder approval; brainstorming; checklists; interviews; constraints analysis; and "qualitative assessment" using probability–impact matrices.

The absence of "risk efficiency" as a basic conceptual framework is my central health warning. Risk efficiency is recognized as a central concept in economics and finance. It is not generally given this central position in project risk management, although it is central to the BP processes mentioned earlier, and related processes used by a number of organizations worldwide, with published discussions dating from the 1970s. Risk efficiency involves maximizing expected reward (possibly by minimizing expected cost) for any given level of risk. Risk here is defined in terms of potential downside departures from expectations, and it is measured (if appropriate) by comparing overlaid cumulative probability distributions for two or three alternative choices between strategies or responses to opportunities or threats. More potential downside variability may mean more risk, but expected values and distribution shapes matter. Sometimes more risk for greater expected reward is appropriate, sometimes it is not, and the difference matters. Risk efficiency is an essential concept for fully understanding how opportunities and threats can be jointly managed in any context. "Risk-reward efficiency" is an alternative term the reader may prefer. What you call it does not matter. Failing to understand it does matter, in direct practical terms.

Linked health warnings include the need to reflect a concern for risk efficiency in the definition of risk and uncertainty adopted; see "uncertainty that matters" as the starting point for understanding risk management, embracing uncertainty in terms of ambiguity as well as variability; understand that "quantitative analysis" in the terms used by this book is bad for your health, as is the usual underlying failure to understand the proper role of risk efficiency, subjective probabilities which may or may not use data, and the irrational behaviour which dependence on hard data and objective probabilities can induce; realize that "quantitative analysis" is not always necessary, but it is usually useful, and it should never be undertaken by those who do not understand it; recognize another version of the 80:20 rule in relation to surveys about best practice or reviews of standards - the 20% minority may be the ones who have got it right.

As a penultimate comment, I strongly agree with the importance of all of David's critical success factors in Chapter 9, but not with some of the details of his responses, because of the linked fundamental differences in conceptual frameworks noted above. But I close by citing from A. A. Milne's classic children's book *Winnie* the *Pooh*: "The Heffalump is a rather large and very important animal. He has been hunted by many individuals using various ingenious trapping devices, but no one so far has succeeded in capturing him. All who claim to have caught sight of him report that he is enormous, but they disagree on his particularities." There is more than a whiff of the Heffalump about the best way to approach processes for managing project opportunities and threats.

Chris Chapman, M.Sc., Ph.D.
Professor of Management Science
University of Southampton
Southampton, United Kingdom

Foreword

Stephen Grey

When project risk management began to emerge as a distinct discipline about fifteen years ago, it was generally greeted as a good idea but it was common to find that no two experts could agree on what it really meant, let alone how it should be addressed. Over the years, a lot of creative energy filtered through a process of trial and error has brought some shape to the subject. There are professional bodies, guidelines, and standards associated with project risk management, and there is even a degree of consensus on the best way to implement it.

However, project risk management is not yet a mature discipline. Project management itself is evolving into a broad-based business process in both the public and the private sector. This means that project risk management has to deal with increasing complexity and issues extending beyond project implementation into strategic planning and business operations while developing links to the growing general risk management and governance requirements of major organizations.

Three aspects of project risk management stand out as the sources of a lot of debate and implementation challenges at the moment:

- The interaction between qualitative risk assessment and quantitative methods
- Integration of risk management across large organizations and through hierarchical management structures

- Dealing with not only uncertainties that have negative consequences, the traditional definition of risk, but also potentially
  advantageous outcomes, often called opportunities

The last of these, dealing with positive and negative consequences of uncertainty, has been the spark of many heated debates. There are those who
would like to confine the term *risk management* to negative consequences,
those who want to consciously integrate risk management with opportunity management, and those, like David Hillson, who believe that they
should be merged. These arguments can be traced back to disagreements
about what to do, the processes being used, and arguments about how to
describe it.

Being more concerned with what to do than how to talk about it, I am
very glad to see this book enter into the debate. David has explored the
language and definitional issues arising when negative and positive sources
of uncertainty are addressed in the one process. For those willing to listen,
he has set out what these issues are and suggested mechanisms for dealing
with them. This should make a significant contribution towards minimizing the arguments about words so that attention can be focused on actions
and process implementation.

It will be interesting to see how the processes of risk management
develop over the next fifteen years and how the language used to talk about
it evolves. Language can be an important element in the development of a
discipline but there is a tendency for debates about definitions to absorb
time and energy far out of proportion to the benefits they produce. This
book should help to resolve some of these issues and move the subject
forward.

Stephen Grey, Ph.D.
Broadleaf Capital International Pty Ltd
Frankston, Victoria, Australia

Preface

Projects are undertaken in order to gain business benefits, involving achievement of objectives in a world characterized by uncertainty. Some of that uncertainty is negative, representing threats to the project's objectives which must be avoided or minimized. But other uncertainties are potentially beneficial opportunities which need to be captured and exploited. While threats can be managed through the routine application of project risk management, there is currently no widely accepted approach for dealing with project opportunities.

This book helps to fill the gap, by extending the familiar threat-based project risk management process to include opportunities explicitly. It provides readers with a clear-thinking rationale and practical techniques for identifying, capturing, and managing opportunities proactively.

Other books on risk management reflect traditional thinking that "risk equals threat." This ground-breaking book is the first to provide a comprehensive and structured framework for opportunity management within projects as an integral part of the risk process. It captures part of the current debate on the future direction of project risk management, unequivocally making the case that definitions of risk must include both opportunity and threat, and risk processes must deal with both equally effectively if project objectives are to be achieved.

The underlying theme of this book is the need for all uncertainty affecting achievement of project objectives to be recognized and managed proactively. Project managers are familiar with the traditional concepts of risk management, and a structured approach to managing risks within projects is increasingly being adopted across a wide range of industries. While precise details of specific methodologies may differ, most project management professionals would recognize a risk process including phases for Definition, Identification, Assessment, Response Planning, and Monitoring, Control, and Update.

There is, however, a systemic weakness in risk management as undertaken on most projects. The standard risk process is limited to dealing only with uncertainties that might have negative impact (threats). This means that risk management as currently practiced is failing to address around half of the potential uncertainties – the ones with positive impact (opportunities).

The implications for project success are clear. Without proactive management of opportunities, the only alternatives are to achieve the plan or to succumb to threats. Risk management becomes a one-way street, with the only option being to travel away from the objectives. However, when opportunities are recognized and addressed effectively, this introduces the "roundabouts and swings" effect where upside risks balance downsides, and it even creates the possibility of improving on the project plan to deliver enhanced benefits.

This book presents the case for extending the traditional risk management process to deal with opportunities explicitly. Starting from an accepted risk process, the reader is introduced to tools and techniques which expose and explore opportunities alongside threats. Project risk management is broadened to address all types of uncertainty, building on the familiar threat-based approach to create an integrated holistic process that includes opportunities. Taking a generic risk management process, each phase is discussed, indicating where modifications are required in order to cover opportunities as well as threats.

The result is an approach to project risk management which reflects the current understanding that risk can be good for you as well as bad. The techniques described in this book will give project management professionals the capability to deal with all types of uncertainty on their projects, both minimizing threats and maximizing opportunities.

All project management professionals will benefit from this book, as well as managers wishing to understand risk management in more depth. The approach is applicable to projects in any industry or country and will prove valuable to students and practitioners alike. Risk management specialists will find the discussion particularly helpful as they seek to resolve the current debate over the definition of risk and the scope of the risk process.

For all readers, the aim is to extend awareness and offer an approach that allows upside uncertainty to be managed, so as to capture as much of the available benefits as possible. By applying the simple process, tools, and techniques described in this book, each reader should be able to prove that "risk can be good for you!" And a new proverb might be coined: "May all your risks be opportunities"

David Hillson, Ph.D.

Acknowledgments

> If I have been able to see further, it was only because I stood on the
> shoulders of giants.
>
> *Sir Isaac Newton (1642–1727)*

Many people have played a part in bringing me to the place where this
book could be written—too many to name individually. But some stand
out as major influences at key milestones in my journey, and it is my
pleasure to give them due credit.

My interest in risk management was initially kindled in the early
1980s at a software metrics conference in Coventry, United Kingdom,
where two keynote presenters spoke with contagious passion about the
importance of risk management. Dr. Robert Charette has remained a
major figure in the risk world, but the untimely death of Paul Rook denied
us his continued insights.

My mentors in project management at Ferranti in the late 1980s,
Keith Jay and Bert Edyvane, both generously encouraged me to explore
the contribution risk management can make to achieving project objectives, and I learned much from their different styles of leadership. I appre-
ciated their willingness to try new approaches, and I believe our projects
benefited from my experimentation.

Former colleagues (now competitors!) at HVR Consulting and
PMProfessional extended my skills in practical application of the risk pro-

cess, notably Chris Thain and Peter Simon, and the mysteries of Monte Carlo simulation were unlocked for me by David Williamson of Euro Log.

I have been privileged to exchange ideas with a number of leading thinkers and practitioners over the years, who have considerably shaped my own thinking and practice. Chief among these gurus are the three other "risk doctors" who have kindly written forewords for this book: Dr. David Hulett, Dr. Chris Chapman, and Dr. Stephen Grey. I value their insights and friendship, and our ongoing constructive debates.

Membership of professional bodies has also provided a wealth of networking opportunities, particularly through the PMI Risk SIG under the inspirational leadership of Chuck Bosler, and the UK APM Risk SIG with its rotating chairmanship. Numerous fellow members have stimulated fruitful new directions for investigation, often unknowingly.

Finally, the most significant influence on my professional and personal development is undoubtedly my wife, Liz. She has my unstinting admiration, respect, and gratitude. During over 25 years of confronting life's uncertainties together, she has shown me how to find the upside, and I am eternally grateful.

To each of these colleagues and others, named and anonymous, I express my thanks for their contribution to my journey into the fascinating world of risk management. Where ideas in this book are derived from the input of another, I have endeavored to provide explicit acknowledgment. Any errors or omissions remain mine alone, in either concept or execution, and are perhaps forgivable as the natural consequences of operating at the leading edge of a fast-developing discipline.

David Hillson, Ph.D.

Contents

10. Future Opportunities!

I

THE CASE FOR OPPORTUNITY MANAGEMENT

1
The Nature of Risk

SOURCES OF UNCERTAINTY

Caius Plinus Secundus, better known as Pliny the Elder (A.D. 23–79), stated in his *Natural History*, "Solum certum nihil esse certi [The only certainty is that there is nothing certain]." In the year before he died, Benjamin Franklin (1706–1790) wrote, "In this life nothing can be said to be certain, except death and taxes." And more recently, Oscar Wilde (1854–1900) declared, "Only the past is certain; the future is at best only probable."

Nowhere is this more evident than in the ever-changing world of business. Today's managers are faced with a bewildering array of uncertainties, as the business environment within which they operate changes at an increasing rate. Perhaps unsurprisingly, many are looking for a solution to this problem, wondering whether it is possible to find a safe path through the fog of an uncertain future. Project management is an attempt to manage this uncertainty, since it is seen as offering a structured approach to produce managed change in a changing environment. The purpose of project management is to act as a change agent, delivering a change to the status quo, and achieving this in a controlled and managed way.

But is project management merely an act of blind faith, trusting to fate or chance in an unpredictable world? Or is there an alternative approach that takes a more responsible attitude to uncertainty? Much of the typical project management process seems to be like trying to drive a car by looking in the rearview mirror, with the concentration on reporting, review, control, and monitoring. Instead projects and businesses need a forward-looking radar, scanning the murky and unclear future to identify the outlines of possible obstacles or shortcuts, allowing the driver to make necessary course corrections in time to avoid disaster and steer toward the desired destination.

The range of uncertainties faced by businesses and their projects today is huge, arising from a multitude of sources including those internal or external to the business, with a range of technical, management, operational, and commercial issues. Some uncertainty is related to the actual work to be done, with the possibility of changing requirements or scope of work, assumptions that may prove to be flawed or even false, use of new technology or novel approaches and methods. Other uncertainties arise from the people involved in the work, including variable skill levels or productivity rates, and the performance of members of the supply chain. Another set of sources are external factors outside the control of the project, including the environment in which the project is undertaken, market conditions, actions of competitors, changing exchange rates or inflation rates, or weather conditions. Then there are the other stake-holders in the project and the business, all of whom by definition are able to influence performance, and may therefore introduce uncertainties into the equation.

Some practitioners have analyzed these sources of uncertainty using hierarchical "breakdown structures." Often these are specific to an industry sector or project type, although it is possible to define generic sources that might affect any type of project in any sector (see Table 1 for a generic example).

Given this huge range of potential uncertainties, one may be tempted to wonder why organizations ever venture into the world of project management. The reason is that there is a clear relationship between uncertainty and risk, and it is a well-recognized and accepted fact that risk is inevitably associated with reward. Most people understand the link between risk and reward, from the simple wager or bet that risks losing the stake to win the prize, to investment in the stock market where prices can fall (the risk) as well as rise (the reward). Similarly, organizations undertake projects specifically to gain the associated benefits, recognizing that they need to manage the inevitable uncertainty to reap the rewards. And the bigger rewards await those who take more risk, as long as they are able to manage it effectively. This is well illustrated by the motto of the famous

Table 1 Sources of Uncertainty for Generic Projects[a]

Level 0	Level 1	Level 2	Level 3
Project risk	Management	Corporate	History/experience/culture
			Organizational stability
			Financial
			Other
		Customer and	History/experience/culture
		stakeholder	Contractual
			Requirements definition
			and stability
			Other
	External	Natural	Physical environment
		environment	Facilities/site
			Local services
			Other
		Cultural	Political
			Legal/regulatory
			Interest groups
			Other
		Economic	Labor market
			Labor conditions
			Financial market
			Other
	Technology	Requirements	Scope uncertainty
			Conditions of use
			Complexity
			Other
		Performance	Technology maturity
			Technology limits
			Other
		Application	Organizational experience
			Personnel skill sets and
			experience
			Physical resources
			Other

[a] After Hall DC, Hulett DT., 2002.

British Army special forces unit, the Special Air Service, or SAS, who fight under the banner "Who dares wins." Without daring to undertake risky projects, organizations will never win the benefits and rewards that those projects can deliver.

UNCERTAINTY AND RISK

Like everything else in life, all projects are inevitably subject to uncertainty. But "uncertainty" does not necessarily mean the same as "risk." How are these two related? The purist statistician or mathematician uses the terms quite differently, although for the man in the street they are regarded as synonyms. The theoretical difference between risk and uncertainty is perhaps best explained by decoding two jargon terms. Risk can be said to be *aleatoric*, whereas uncertainty is described as *epistemic*.

> *Alleatoric* is derived from the Latin word *alea*, meaning dice. This indicates that a risk is an event where the set of possible outcomes is known, and the probability of obtaining each outcome can be measured or estimated, but the precise outcome in any particular instance is not known in advance. Thus "risk" strictly refers to an unknown event drawn from a known set of possible outcomes.
>
> *Epistemic* comes from the Greek word *episteme*, meaning knowledge. The suggestion here is that uncertainty relates to a lack of knowledge about possible outcomes, including both their nature and associated probabilities. An "uncertainty" is thus an unknown event from an unknown set of possible outcomes.

The relationship between risk and uncertainty, and the distinction between *aleatoric* and *epistemic*, are captured in the following couplet:

> Risk is measurable uncertainty;
> Uncertainty is unmeasurable risk.

Another way of distinguishing these related terms of "risk" and "uncertainty" is to consider the relationship between knowledge and awareness, as illustrated in Figure 1 (this is similar in concept to the Johari Window, which describes self-awareness and perception by others). In the top right quadrant of Figure 1 lies *Certainty*, where both knowledge and awareness are present, and the extent of what is known is fully understood. Below this is the area of *Amnesia*, where there is no aware-

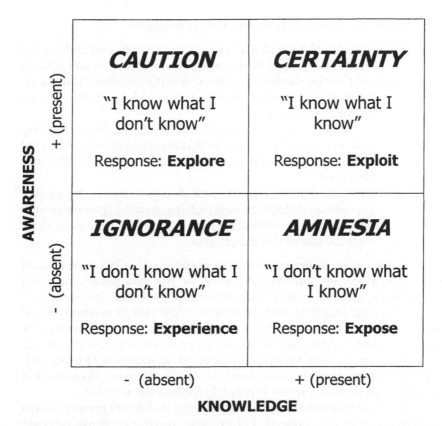

AWARENESS

+ (present)

CAUTION

"I know what I don't know"

Response: **Explore**

CERTAINTY

"I know what I know"

Response: **Exploit**

– (absent)

IGNORANCE

"I don't know what I don't know"

Response: **Experience**

AMNESIA

"I don't know what I know"

Response: **Expose**

– (absent)   + (present)

KNOWLEDGE

Figure 1 Knowledge and awareness.

ness or a blind spot regarding the knowledge actually possessed. The top left corner requires *Caution*, arising from being aware of an absence of knowledge, understanding the scope of the shortfall. And "unknown unknowns" reside in the bottom left section, with pure *Ignorance* of the situation faced.

In terms of aleatoric risk and epistemic uncertainty, we find risk in the Caution zone, with known unknowns, i.e., events and circumstances where we are aware that we do not have all the necessary facts. Uncertainty occupies the bottom two quadrants, including both Amnesia and Ignorance. (There is clearly neither risk nor uncertainty associated with the Certainty zone!)

Each zone warrants a different type of response:

Certainty should be *exploited*, playing to strengths and making full use of known facts to take well-founded decisions and actions.

Areas of *Caution* should be *explored*, seeking to understand the aspects where there is known to be a knowledge deficit or a recognized weakness.

Amnesia needs to be *exposed*, probably through a facilitated process, to unlock the knowledge that exists and allow it to be used effectively, avoiding wasted opportunities that might otherwise be missed.

Ignorance can only be tackled through *experiment* or by gaining *experience* (perhaps through training, mentoring, or on-the-job application), growing in both knowledge and awareness, to reduce the size of this danger area.

Clearly any risk management process must address these last three types of response, exploring and exposing areas of risk and uncertainty, and providing a means of gaining useful experience to counter ignorance.

It is clear therefore that the technical specialist in mathematics or statistics will use the terms "uncertainty" and "risk" quite differently. For businesses and projects, there is also a difference between these two, which is more pragmatic than the abstruse technical definitions used by specialists. It is important to understand and clarify this difference, to ensure that any risk management process is properly focused and targeted.

It is clear that there are some uncertainties that do not present risks to a given business or its projects. For example, variable exchange rates are irrelevant to an organization operating entirely within one currency regime. Weather conditions do not usually affect projects undertaken in an indoors office environment. Possible changes in regulatory frameworks are of little interest to businesses whose projects are outside the scope of those regulations. Clearly only a subset of all uncertainties are relevant as risks to a particular business or project. What determines which subset of uncertainties qualify as risks?

The key factor in transforming uncertainties into risks is the fact that organizations and projects have specified *objectives* to be met. For the organization, these objectives are the strategic aims of the business, whereas project objectives are defined in the project charter or business case. The objectives define the measures of success, and are usually expressed in fixed measurable terms—certain levels of profitability, rate of return, or market share, or meeting milestones, budget, and performance targets. Of all the possible future states we might imagine as possible, the objectives describe our desired future, our goal or mission, the end to which our endeavors are reaching.

In project management, these objectives are most often expressed as a combination of time, cost, and quality/performance/scope, creating an "iron triangle" of fixed success criteria against which the project is measured. Yet all experienced project managers know that reality is not likely to be as simple as defining the objectives then meeting them. Trade-offs are common: as the brochure of one business puts it, "We offer low prices, rapid delivery, high quality—choose any two!"

Why do businesses and projects have difficulty in meeting defined objectives? Because, as Plato (427–347 B.C.) realized many centuries ago, "The problem with the future is that more things might happen than will happen." Science fiction writers have recognized this fact and used the device of parallel universes to explore alternative futures. Out of the almost infinite combination of events, circumstances, and conditions that might exist, only one specific combination will actually occur, and this one potential future will become reality, passing through the present into history. The problem for business planners and project managers is that we don't know today which one of the many possible futures will actually come to pass. Danish Nobel Prize-winning physicist Niels Bohr (1885–1962) recognized this when he said, "Prediction is very difficult, especially about the future."

This is illustrated in Figure 2, which shows a wide range of possible futures radiating out from the present moment in time. Some of these futures would be welcome, as they would meet our objectives and aspirations to a greater or lesser extent. Other futures would be less welcome, representing adverse circumstances that we would not wish to see. And some futures are currently unknown and/or unknowable, since they are dependent on combinations of events or conditions that do not exist.

Faced with this range of futures, we are constantly required to make choices, decisions, or actions (which also include refusals to choose, or deferral of deciding, or deliberate inaction). Each choice, decision, or action results in some of the set of possible futures ceasing to exist, as they become infeasible in the light of the new changed circumstances. But additional possible futures are also created by the choice, decision, or action, resulting in a new range of options to be considered. At each moment the range of possible futures is different from that which existed immediately prior to the choice, and different from all other ranges at any other time in the past. (The scenario-planning technique by which such

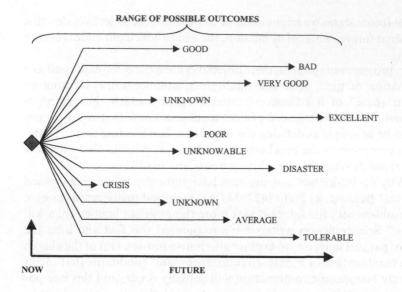

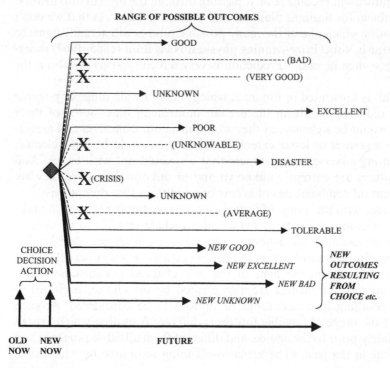

Figure 2 Future outcomes.

alternative futures are identified and the effect of decisions is evaluated is known as Field Anomaly Relaxation, or FAR, and is widely used in strategic decision-making support.)

The challenge for the business and project management team is consistently to make choices, decisions, and actions that move toward the desired objectives—a task that is made difficult by the ever-changing profile of uncertain futures. The only thing that can be said with confidence about the future is that it is full of uncertainty (although a skeptic may be tempted to insist that the best we can say is "Everything about the future is uncertain—probably!").

There are a great many uncertainties facing businesses and projects, as we have seen above. Some of these have the potential to affect objectives, and it is this interaction between uncertainty and objectives that creates risk. Risk cannot exist in a vacuum; there must be a risk to something. Objectives define what is "at risk" from the potential effect of uncertainty, and these two factors must both be present to give rise to risk. Figure 3 shows the relationship between these two, illustrating that it is *uncertainty*

Risk arises from interactions between
• *objectives* ...
• what *must* happen
• *uncertainty* ...
• what *might* happen

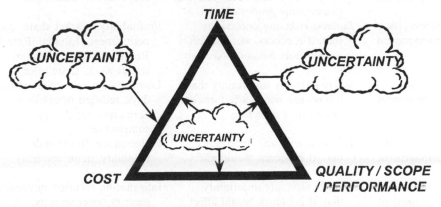

Figure 3   Risk results from uncertainty operating on objectives.

(whether internal or external, arising from technical, management, or commercial sources) operating on *objectives* that causes risk. If there were no uncertainty and the future was perfectly knowable and adequately known, there would be no risk. Similarly, if objectives were flexible rather than fixed, and could be varied to cope with the effects of uncertainty, risk would not exist.

It is the interaction of uncertainty on objectives that gives rise to risk. This allows us to determine which uncertainties are relevant to a business or project: only those uncertainties that have the potential to affect objectives can become risks. A risk is an uncertainty that matters, and "mattering" is defined in relation to specific objectives.

In other words, *risk* can be defined as *any uncertainty that, if it occurs, would affect one or more objectives*.

This definition allows the term "risk" to be used very widely in a variety of applications. Different types of risk are distinguished by the types of objectives affected, as illustrated in Table 2. The most common

Table 2 Relationship Between Risk and Objectives

Type of risk management	Definition	Sample objectives
Generic	*Risk*: any uncertainty that, if it occurs, would affect one or more *objectives*	–
Project risk management	*Project risk*: any uncertainty that, if it occurs, would affect one or more *project objectives*	Time, cost, performance, quality, scope, client satisfaction
Business risk management	*Business risk*: any uncertainty that, if it occurs, would affect one or more *business objectives*	Profitability, market share, competitiveness, Internal Rate of Return (IRR), reputation, repeat work, share price
Safety risk management	*Safety risk*: any uncertainty that, if it occurs, would affect one or more *safety objectives*	Low accident rate, minimal lost days, reduced insurance premiums, regulatory compliance
Technical risk management	*Technical risk*: any uncertainty that, if it occurs, would affect one or more *technical objectives*	Performance, functionality, reliability, maintainability
Security risk management	*Security risk*: any uncertainty that, if it occurs, would affect one or more *security objectives*	Information security, physical security, asset security, personnel security

application of risk management is in projects, where project risks are defined as those uncertainties that could affect project objectives. In a similar way, we can define business risks as those uncertainties facing business objectives; strategic risks relate to strategic objectives; reputational risk management addresses uncertainties that could affect the organization's standing or profile. The same is true of operational risk, technical risk, environmental risk, safety risk, security risk, information risk, business continuity risk, health risk, country risk, market risk, etc. The key is first to define the objectives, after which risks can be identified and managed.

PROJECTS AND RISK

A project can be defined as "a unique and temporary endeavor that introduces change to create a product or service that meets defined objectives using various resources within set constraints." Everyone agrees that projects are inherently risky, and the "zero-risk project" does not exist. So what makes projects risky?

Some characteristics are built into the very fabric of all projects, which make projects risky by nature, including:

Uniqueness, involving at least some elements that have not been done before

Deliverables, whether product or service, which the project must create to deliver the benefits for which it was undertaken, and which may not perform as expected or required

Assumptions, both implicit and explicit, about various aspects of the project and its environment, which may prove to be unfounded or changeable

Constraints and objectives, defining the measures by which project success will be determined, which are usually fixed and sometimes conflicting

Different stakeholders with requirements, expectations, and objectives that are varying, overlapping, and sometimes conflicting

People, including project team members and management, clients and customers, suppliers and subcontractors, all of whom are unpredictable to some extent, with variable productivity, changing attitudes and motivations, and developing relationships and interfaces

Change, both as a verb (a project is a change agent) and as a noun (a project creates a changed situation), which necessarily involves moving into an unknown future

Environment within which the project exists, including both the business and organizational environment where changes in strategy may affect project priority or resourcing, and the wider environment influenced by public, political, social, ethical, and regulatory factors, usually outside the project's control

These characteristics are intrinsic to the nature of a project, and cannot be changed without significantly affecting the project itself. For example, a "project" that was not unique, had no constraints, involved no people, and did not introduce change would in fact not be a project at all. Trying to remove the risky elements from a project would turn it into something that could not be described as a project. Indeed projects are undertaken to gain benefits while taking the associated risks in a controlled manner. If, as discussed above, risk arises from the interaction of uncertainty on objectives, it is impossible to conceive a worthwhile project that had either no uncertainty or no objectives, and so it is equally impossible to conceive a nonrisk project.

As a direct result of the inherent risky nature of projects, it is not possible to have a project that is not exposed to risk. Of course some projects will be high risk, while others are less risky, but all projects are by definition risky to some extent. The "zero-risk project" is an oxymoron, a meaningless phrase describing something that cannot by definition exist. This of course is why risk management is such an important part of effective project management: since all projects are exposed to risk, successful projects are the ones where that risk is properly managed.

RISK AND OPPORTUNITY

We have seen that risk arises from the influence of uncertainty on objectives. There is, however, a question over the nature of such an effect. Is the effect of a risk always and wholly negative? This question is the subject of current debate among risk practitioners, which has profound implications for the practice of risk management. The debate is explored in more detail in Chapter 2 and the Appendix, but it is outlined here in summary.

There is no doubt that businesses and projects, like everything else in life, are subject to uncertainty. It is also clear that if and when uncertainty strikes, it can have a range of effects on achievement of objectives, from the total disaster to the unexpected welcome surprise. Some uncertainty might be harmful if it came to pass (we could call such uncertainties "threats"), whereas other uncertainties might assist in achieving our objectives (these might be known as "opportunities"). It is important to understand the relationship between threats and opportunities, especially in the context of projects and project risk management.

Some people claim that opportunities are simply the reverse side of threats, and that it is possible to transform all threats into opportunities by using clever semantics to turn a negative statement into a positive. This suggests that opportunities do not exist in their own right, but are merely the inverse or absence of identified threats. For example, the threat that "The contractor may be late in delivering the agreed equipment" might be expressed as an opportunity that "The contractor may not be late in delivering the agreed equipment." Or the threat that "Interest rates might go up" can be balanced with the opportunity that "Interest rates might come down." This type of trivial wordplay does not represent the possibility of identifying real opportunities separate from threats, and is probably a waste of time.

A standard examination question for first-year UK medical students requires them to write an essay on the following topic: "Health is not the absence of disease—discuss." A similar approach could be taken to the issue of opportunities and threats, recognizing that "Opportunity is not the absence of threat." Distinctive opportunities exist in their own right, presenting the chance to enhance project objectives, deliver early, cost less, increase customer satisfaction, improve competitiveness, enhance company reputation, etc. Viewing opportunities simply as the inverse or absence of threats is likely to result in some real opportunities being missed, with subsequent loss to the project of the chance to improve performance.

Indeed, if opportunity is only defined as the absence of threat, then risk management becomes a one-way street, with the only option being to meet the plan or miss it by a greater or lesser extent, depending on how many threats occur. Denying opportunities to improve means losing the upside, leaving only the possibility of downside impacts. The only question in this situation is how bad the failure will be.

The reality is that, having set a realistic plan for achieving project objectives, the uncertain environment in which the project is executed will

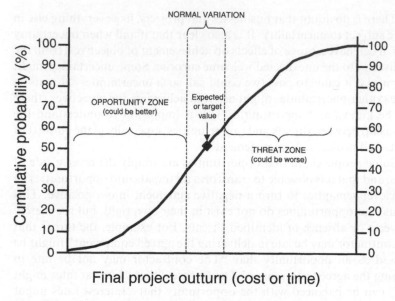

Figure 4   Threat and opportunity in project outturns.

produce a range of possible impacts, including both upside opportunities and downside threats, as illustrated in Figure 4. Management might choose where to set the expected or target value, and can determine the range of variation to be considered "normal" for this type of project, but it will always be true that variation outside this range is possible, on both sides. As a result, opportunities to do better than plan will exist, as well as threats to do worse.

There is no doubt that a good project manager will deal with opportunities proactively on his or her project, and that project management must encompass opportunity management as well as dealing with threats. There are two equal failings that any project manager must avoid. The first is if a threat occurs on the project that could have (and should have) been foreseen and that could have been avoided if action had been taken earlier. This represents a clear failure of the risk management process. But there is a second failing that is equally embarrassing and unprofessional for a project manager, namely to miss an opportunity that could have (and should have) been foreseen and could have been exploited or realized if action had been taken earlier. The occurrence of avoidable threats and the nonoccurrence of achievable opportunities should be viewed as equally bad outcomes, and the professional project manager should take steps to ensure that both threats and opportunities are identified sufficiently in advance to allow them to be managed effectively.

The issue is therefore not whether opportunities should be managed as well as threats. It is rather whether we could or should include both types of uncertainty in our definition of "risk," and whether both could or should be handled by a common "risk management process."

ONE DEFINITION OR TWO?

There is currently a vigorous debate among the risk management community, with individuals and groups taking and defending strong opposing positions. The issue is whether the term "risk" should encompass both opportunities and threats, or whether "risk" is exclusively negative, with "opportunity" being qualitatively distinct. There appear to be two options:

1. "Risk" is an umbrella term, with two varieties:
   "Opportunity," which is a risk with positive effects
   "Threat," which is a risk with negative effects
2. "Uncertainty" is the overarching term, with two varieties:
   "Risk" referring exclusively to a threat, i.e., an uncertainty with negative effects
3. "Opportunity," which is an uncertainty with positive effects

There is no doubt that common usage of the word "risk" sees only the downside. Asking the man in the street if he would like to have a risk happen to him will nearly always result in a negative response—"Risk is bad for you." This is reflected in the traditional definitions of the word, both in standard dictionaries and in some technical definitions.

However, some professional bodies and standards organizations have gradually developed their definitions of "risk" to include both upside and downside. Several of these have neutral definitions where the nature of the effect is undefined, and which could therefore implicitly encompass both positive and negative effects. Others are explicit in naming both opportunities and threats within their definition of "risk." The evolution of these "official" definitions of risk is discussed in Chapter 2.

One might ask whether this matters, since as William Shakespeare (1564–1616) said, "That which we call a rose, by any other name would smell as sweet." It is important, however, because the decision to encompass both opportunities and threats within a single definition of risk would be a clear statement of intent, recognizing that both are equally important influences over business and project success, and accepting that both need managing proactively.

Opportunities and threats are not qualitatively different in nature, since both involve uncertainty, which has the potential to affect objectives. An opportunity can be defined as an uncertain event or set of conditions that, if it occurs, would benefit the project or business. Similarly, we might define a threat as an uncertain event or condition that, if it occurs, would harm the project or business. Both definitions are the same, apart from the type of effect on objectives. As a result, it makes sense to bring the two together under a common definition that combines the uncertainty element with the potential to affect objectives, and that is precisely how "risk" is defined (as illustrated in Fig. 5).

As a result, both can be handled by the same process, although some modifications may be required to the standard risk management approach to enable it to deal effectively with opportunities.

Risk: any uncertainty that, if it occurs, would affect one or more objectives	
Threat:	Opportunity:
any uncertainty that, if it occurs, would affect one or more objectives negatively	any uncertainty that, if it occurs, would affect one or more objectives positively

Figure 5   Relationship between risk, threat, and opportunity.

ONE PROCESS OR TWO?

Linked to the discussion about definitions of risk is a parallel debate about processes. Those who define "risk" as wholly negative and who see "opportunity" as something distinct naturally advocate separate processes for risk management and opportunity management. Conversely, those who view "risk" as a common term encompassing both opportunities and threats accept the possibility of managing both in an integrated manner through a common process. For example, the *Guide to the Project Management Body of Knowledge* from the Project Management Institute (PMI PMBoK) defines risk management as "The systematic process of identifying, analysing, and responding to project risk. It includes maximising the probability and consequences of positive events and minimising the probability and consequences of negative events to project objectives."

Despite this clear scope, the risk management process described in the PMI PMBoK still tends to focus on management of threats, reflecting the common experience of risk practitioners who find it easier to identify potential pitfalls and problems than to look for hidden advantages or upsides. Other risk management processes that claim to recognize both positive and negative risks also appear to pay similar lip service to opportunity management, failing to match their broad inclusive definition with a process that copes explicitly with both types of risk.

In addition to the theoretical point that if opportunity and threat are two varieties of the same thing, they should be managed together, the use of a common process to deal with both types of risk together under "risk management" has several practical advantages.

The first is simply to ensure that opportunities are indeed identified and managed. Including opportunities in the definition of risk will mean that there is no need to introduce a new process, with all the resistance that is likely to be encountered when another overhead process is added. If a separate "opportunity management" process is required, this is likely to be seen by hard-pressed project managers as an additional burden, and as such it may not receive appropriate attention. If, on the other hand, opportunities are handled by a risk management process that already exists, the additional overhead is minimized. This also leads to increased efficiency—a single process dealing with two types of issue will be more efficient than two separate processes.

Second, the risk management mind-set encourages identification of uncertainties that might affect objectives, and leads the project manager to look for ways of addressing these proactively. While the need to spot and neutralize or minimize threats is clear, it is less natural to spend time looking out for unplanned opportunities. Most project managers think of "risk" in terms of threats alone. A broader risk management mind-set that includes opportunities encourages the identification of all uncertainties that might affect objectives, and leads the project manager to look for ways of capturing risk proactively, combating the one-sided approach as commonly practiced. A common process can deal with positive uncertainties in the same way as negative ones, extending the familiar risk management approach.

However, if the existing risk management process is to be extended to allow opportunities to be managed alongside threats, some changes will be required. Part II of this book suggests where such modifications should be focused.

INITIAL CONCLUSIONS

We have seen that life, business, and projects are subject to uncertainty. For projects this uncertainty is inherent in the nature of the project, and it is also desirable since uncertainty is closely related to reward. Uncertainty becomes risk through its interaction with objectives, with a risk being defined as any uncertainty that, if it occurs, would affect one or more objectives. However, this effect can be either positive or negative, leading to the suggestion that the term "risk" could encompass both opportunities (uncertainties with positive effects on objectives) and threats (uncertainties with negative effects on objectives). There are clear benefits in a common process to handle both types of uncertainty, upside as well as downside, although this is likely to require some changes to the current approach to risk management.

Chapter 2 examines how risk management is usually practiced within projects, and lays the ground for development of a wider process that can deal proactively and effectively with both opportunities and threats.

REFERENCES

Artto, K. A., Kähkönen, K., Pitkanen, P. J. (2000). *Unknown Soldier Revisited: A Story of Risk Management*. Helsinki: PMA Finland.

Baker, R. W. (1986). Handling uncertainty. *Int. J. Project Management* 4(4):205–210.

Bartlett, P. L. (2002). *Managing Risk for Projects and Programmes: A Risk Handbook*. Hook, Hampshire, UK: Project Manager Today Publications.

Bernstein, P. L. (1996). *Against the Gods—The Remarkable Story of Risk*. New York: John Wiley.

Borge, D. (2001). *The Book of Risk*. New York: John Wiley.

CCTA. (1993). *Introduction to the Management of Risk*. London: HMSO.

CCTA. (1994). *Management of Project Risk*. London: HMSO.

Chapman, C. B., Ward, S. C. (1997). *Project Risk Management: Processes, Techniques and Insights*. Chichester, UK: John Wiley.

Chapman, C. B., Ward, S. C. (2000). Estimation and evaluation of uncertainty—a minimalist first-pass approach. *Int. J. Project Management* 18(6):369–383.

Chapman, C. B., Ward, S. C. (2002). *Managing Project Risk and Uncertainty*. Chichester, UK: John Wiley.

Chicken, J. C., Posner, T. (1998). *The Philosophy of Risk*. London: Thomas Telford.

Courtney, H., Kirkland, J., Vignerie, P. (1997). Strategy under uncertainty. *Harvard Business Rev.* 75(6):66–79.

Davis, M. (2001). What is risk? It's important we get it right. *InfoRM, J. UK Inst. Risk Management* June/July:16–19.

de Cano, A., de la Cruz, M. P. (1998). The past, present and future of project risk management. *Int. J. Project Business Risk Management* 2(4):361–387.

Dembo, R. S., Freeman, A. (1998). *Seeing Tomorrow—Rewriting the Rules of Risk*. New York: John Wiley.

Dorofee, A. J., et al. (1996). *Continuous Risk Management Guidebook*. Pittsburgh, PA: Software Engineering Institute, Carnegie Mellon University.

Fischnoff, B., Watson, S. R., Hope, C. (1984). Defining risk. *Policy Sci.* 17:123–139.

Flyvbjerg, B., Bruzelius, N., Rothengatter, W. (2003). *Megaprojects and Risk: An Anatomy of Ambition*. Cambridge, UK: Cambridge University Press.

Hall, D. C., Hulett, D. T. (2002) *Universal Risk Project—Final Report*. Available from PMI Risk SIG website *www.risksig.com/articles/UR%20Project%20Report.doc*.

Hillson, D. A. (1998). Managing risk. *IEE Rev.* 44(1):31.

Hillson, D. A. (June/July 1999). Business uncertainty: threat or opportunity? *ETHOS Mag.* 14–17 (Issue 13).

Hillson, D. A. (1999). Managing risk—the critical factor in successful project management. *Cost Engin.* 37(1):11–12.

Hillson, D. A. (July–December 2002). Critical success factors for effective risk management. (four-part series). *PM Rev.*

Hillson, D. A. (2002). Extending the risk process to manage opportunities. *Int. J. Project Management* 20(3):235–240 [Also in *Proceedings of the 4th Euro-*

pean Project Management Conference (PMI Europe 2001), presented in London, UK, 6–7 June 2001.]

Hillson, D. A. (2002). Extending the risk process to manage opportunities. *Projects, Profits* 2(11):19–24.

Hillson, D. A. (April 2002). What is risk? Towards a common definition. *InfoRM, J. UK Inst. Risk Management* 2:11–12.

Hillson, D. A. (2003). A little risk is a good thing. *Project Manager Today* 15(3):23.

HM Government Cabinet Office Strategy Unit. (2002). Risk: Improving government's capability to handle risk and uncertainty. Report ref 254205/1102/ D16. Crown Copyright.

Hulett, D. T., Hillson, D. A., Kohl, R. (2002). Defining risk: a debate. *Cutter IT J.* 15(2):4–10.

Institution of Civil Engineers (ICE) and Faculty, Institute of Actuaries. (1997). *Risk Analysis, Management for Projects (RAMP)*. London: Thomas Telford.

Jaafari, A. (2001). Management of risks, uncertainties and opportunities on projects: time for a fundamental shift. *Int. J. Project Management* 19(2): 89–101.

Kähkönen, K., Artto, K. A., eds. *Managing Risks in Projects*. London: E,FN Spon.

Kähkönen, K. (2001). Integration of risk and opportunity thinking in projects. Proceedings of the 4th European Project Management Conference (PMI Europe 2001), presented in London, UK, 6–7 June 2001.

Kliem, R. L., Ludin, I. S. (1997). Reducing Project Risk. Aldershot, UK: Gower.

Luft, J. (1961). The Johari window. *Human Relations Training News* 5(1):6–7.

Luft, J. (1969). *Of human interaction*. Palo Alto, CA: National Press Books.

Milliken, F. J. (1987). Three types of perceived uncertainty about the environment: state, effect and response uncertainty. *Acad. Management Rev.* 12(1):133–143.

Morris, P. W. G., Hough, G. H. (1987). *The Anatomy of Major Projects: A Study of the Reality of Project Management*. Chichester, UK: John Wiley.

Newland, K. E. (1997). Benefits of project risk management to an organisation. *Int. J. Project Business Risk Mgt.* 1(1):1–14.

O'Reilly, P. (1998). *Harnessing the Unicorn—How to Create Opportunity and Manage Risk*. Aldershot, UK: Gower.

Pender, S. (2001). Managing incomplete knowledge: why risk management is not sufficient. *Int. J. Project Management* 19(2):79–87.

Project Management Institute. *A Guide to the Project Management Body of Knowledge (PMBoK®)*. 2000 edi. Philadelphia: Project Management Institute.

Rosa, E. A. (1998). Metatheoretical foundations for post-normal risk. *J. Risk Res.* 1(1):15–44.

Simon, P. W., Hillson, D. A., Newland, K. E., eds. *Project Risk Analysis, Management (PRAM) Guide*. High Wycombe, Bucks, UK: APM Group.

UK Association for Project Management. (2000). *Project Management Body of Knowledge* 4th ed. High Wycombe, Bucks, UK: APM.

Ward, S. C., Chapman, C. B. (2003). Transforming project risk management into project uncertainty management. *Int. J. Project Management* 21(2):97–105.

Ward, S. C. (2001). Project uncertainty management as a desirable future. In: *Proceedings of the 4th European Project Management Conference* (PMI Europe 2001), presented in London, UK, 6–7 June.

Ward, S. C., Klein, J. H., Avison, D. E., Powell, P. L., Keen, J. (1997). Flexibility and the management of uncertainty: a risk management perspective. *Int. J. Project Business Risk Management* 1(2):131–145.

2
Existing Approaches to Risk Management

EVOLVING UNDERSTANDING OF RISK

Risk management practitioners do not agree about the definition of risk. Given the long history of risk management it might be surprising that this question still excites any interest at all. But risk management, like all disciplines, is not standing still, and the definition question is part of the ongoing development of risk management as an essential tool for the effective manager.

One key aspect of the definition debate (introduced in Chapter 1) centers on the question of whether the term "risk" should only be used to refer to uncertain events that could have an effect that is unwelcome, adverse, negative, or harmful (i.e., threats). The common usage of the word is associated with these types of effect, expressed in the concept that "Risk is bad for you." However, there is another view that is held by an increasing number of risk practitioners, especially in the project risk management field, that risk management should address uncertainties with positive impacts (opportunities) as well as those threats traditionally covered by the process. This has led to a perspective that the term "risk" should include both threats and opportunities.

Some risk practitioners strongly oppose this move, while others feel that it is an essential forward step. Each side of this debate claims support from the silent majority, either stating that nobody wants to change the definition from threat-only or asserting that there is a growing appetite for change. Neither of these statements currently has any objective supporting evidence, both being based largely on anecdotal data. Results from a recent survey are presented and discussed in the Appendix to test the strength of feeling on both sides of the discussion.

Of course, it is possible to find respected and authoritative risk management writers and practitioners who take strong positions on either side of this debate. Different organizations also use approaches to risk management based either on narrow definitions of risk that are threat-only or on broader definitions including opportunity. However, rather than counting the number of individuals, experts, or organizations who use one approach or another, it is perhaps more interesting and relevant to review the position taken by the various risk management standards documents that currently exist or have been recently published.

"STANDARD DEFINITIONS" OF RISK

Clearly, if risk management is to be used effectively, risk management professionals should agree on their use of the terminology. This means clarifying whether or not risk includes upsides (opportunities) as well as downsides (threats). Many different professional bodies and standards institutions have attempted to create a definition of risk that will be widely accepted. Of course, these perspectives differ, but there seems to be a trend toward a common view, as examination of some key sources reveals.

The traditional standard English dictionaries are agreed that the definition of "risk" is exclusively negative, and is a synonym for "threat." But professional bodies take a different view from the layman, as is often the case with technical language, which may have a special meaning within a discipline, and which can therefore differ from colloquial usage.

Some professional risk management standards and guidelines follow the traditional dictionary definition by promoting a wholly *negative* view of risk. Others adopt a *neutral* definition where the effect of risk is undefined, thus allowing implicitly both upside and downside impacts within the definition. Finally, another group of standards take a *broader* approach, explicitly including opportunity as part of risk.

The following sections detail these three groups in turn, and Table 1 provides a summary of the definitions in each group, indicating the breadth of the debate. This review does not presume to be complete or comprehensive, since there are surely other national and international risk management standards that have been inadvertently omitted. The review does, however, include all those risk management publications of which the author was aware at the time of writing, and the standards in this section were not included selectively to make a particular point.

Negative Definitions

National standard-setting bodies equating "risk" with "threat" include the first official risk management standard NS5814:1991 *Krav til risiko-analyser* published by Norges Standardiseringsforbund in Norway, which defined risk as "the danger that undesirable events represent."

The international standard IEC 300-3-9:1995 and its British Standard equivalent BS8444-3:1996 *Guide to Risk Analysis of Technological Systems* also limit risk to something negative, defining it as "combination of the frequency or probability of occurrence and the consequence of a specified hazardous event."

The Canadian Standards Association's CAN/CSA-Q850-97:1997 *Risk Management Guideline for Decision-Makers* speaks of risk as "the chance of injury or loss."

The UK Construction Industry Research and Information Association (CIRIA) take a similarly negative view in their 1996 *Guide to the Systematic Management of Risk from Construction*, with risk being described as "chance of an adverse event."

Also in the United Kingdom, the government agency responsible for IT project management (Central Computer and Telecommunications Agency, CCTA) published its *Managing Successful Programmes (MSP)* guidance in 1999, which defines programme risks as "events or situations that may adversely affect the direction of the programme, the delivery of its outputs or achievement of its benefits."

In the United States, the third edition (January 2000) of the US Department of Defense (DoD) DSMC (Defense Systems Management College) *Risk Management Guide for DoD Acquisition* defines risk as "a measure of the potential inability to achieve overall program objectives within defined cost, schedule and technical constraints," and this definition has persisted into the fourth and fifth editions.

Finally in this group, and also originating in the United States, the Institute of Electrical and Electronics Engineers (IEEE) have recently issued IEEE 1540:2001 *Software Life Cycle Processes—Risk Management* covering risk management in the software engineering context, which de-

Table 1 Definitions of "Risk" in Standards and Guidelines

Negative definitions	Neutral definitions	Broad definitions
Norges Standardiseringsforbund NS5814:1991 "the *danger* that *undesirable events* represent"	UK Association for Project Management PRAM Guide 1997 "an uncertain event or set of circumstances which, should it occur, will have *an effect* on achievement of ... objectives"	British Standard BS ISO 10006:1997 "potential *negative events and ... opportunities for improvement* ... the term risk covers *both aspects*"
IEC 300-3-9:1995, and British Standard BS8444-3:1996 "combination of the frequency ... of occurrence and the consequence of a specified *hazardous event*"	Standards Australia/Standards New Zealand AS/NZS 4360:1999 "the chance of something happening that will have *an impact* upon objectives"	UK Institution of Civil Engineers RAMP Guide 1997 "a *threat (or opportunity)* which *could affect adversely (or favorably)* achievement of the objectives"
UK Construction Industry Research & Information Association (CIRIA) 1996 "chance of an *adverse event*"	British Standard BS6079-3:2000 "uncertainty ... that *can affect* the prospects of achieving ... goals"	British Standard BS6079-1:2002 and BS6079-2:2000 "combination of the probability ... of a *defined threat or opportunity* and the magnitude of the consequence"
Canadian Standards Association CAN/CSA-Q850-97:1997 "the chance of *injury or loss*"	British Standard PD 6668:2000 "chance of something happening that will have *an impact* upon objectives"	Project Management Institute PMBoK 2000 "an uncertain event or condition that, if it occurs, has a *positive or negative effect* on a project objective ... includes both *threats* to the project's objectives *and opportunities* to improve on those objectives"
UK CCTA MSP 1999 "events or situations that may *adversely affect* the direction of the programme, the delivery of its outputs or achievement of its benefits"	British Standard BS IEC 62198:2001 "combination of the probability of an event occurring and its *consequences* for project objectives"	British Standard BSI PD ISO/IEC Guide 73:2002, and UK Institute of Risk Management 2002 "combination of the probability of an event and its consequence ... consequences can range from *positive to negative*"
US DoD DSMC 2000 "potential *inability to achieve* overall program objectives"		UK Office of Government Commerce MoR 2002 "uncertainty of outcome, whether *positive opportunity or negative threat*"
IEEE 1540:2001 "the likelihood of an event, *hazard, threat* or situation occurring and its *undesirable consequences*; a *potential problem*"		UK MoD Risk Management Guidance 2002 "a significant uncertain occurrence ... defined by the combination of the probability of an event occurring and its consequences on objectives ... the term 'risk' is generally used to embrace the possibility of *both negative and/or positive consequences*"

Source:

British Standards Institute. 1996. British Standard BS8444-3:1996 (previously issued as IEC 300-3-9:1995) "*Risk Management: Part 3—Guide to Risk Analysis of Technological Systems.*" British Standards Institute, London, UK. ISBN 0-580-26110-7.

British Standards Institute. 1997. British Standard BS ISO 10006:1997 "*Quality Management—Guidelines to Quality in Project Management.*" British Standards Institute, London, UK. ISBN 0-580-28801-3.

British Standards Institute. 2000a. British Standard BS6079-2:2000 "*Project Management—Part 2: Vocabulary.*" British Standards Institute, London, UK. ISBN 0-580-33148-2.

British Standards Institute. 2000b. British Standard BS6079-3:2000 "*Project Management—Part 3: Guide to the Management of Business-related Project Risk.*" British Standards Institute, London, UK. ISBN 0-580-33122-9.

British Standards Institute. 2000c. BSI PD 6668:2000 "*Managing Risk for Corporate Governance.*" British Standards Institute, London, UK. ISBN 0580-33246-2.

British Standards Institute. 2001. BS IEC 62198:2001 "*Project Risk Management—Application Guidelines.*" British Standards Institute, London, UK. ISBN 0-580-390195.

British Standards Institute. 2002a. British Standard BS6079-1:2002 "*Project Management—Part 1: Guide to Project Management.*" British Standards Institute, London, UK. ISBN 0-580-39716-5.

British Standards Institute. 2002b. BSI PD ISO/IEC Guide 73:2002 "*Risk Management—Vocabulary—Guidelines for Use in Standards.*" British Standards Institute, London, UK. ISBN 0-580-401782.

Canadian Standards Association. 1997. National Standard of Canada CAN/CSA-Q850-97 "*Risk Management: Guideline for Decision-makers.*" Canadian Standards Association, Ontario, Canada. ISSN 0317-5669.

Godfrey, P. 1996. "*Control of Risk: A Guide to the Systematic Management of Risk from Construction.*" Construction Industry Research & Information Association (CIRIA), London, UK. ISBN 0-86017-441-7.

Institute of Electrical and Electronics Engineers (IEEE). 2001. IEEE 1540:2001 "*Software Life Cycle Processes—Risk Management.*" IEEE, New York.

Institution of Civil Engineers (ICE) and Faculty and Institute of Actuaries. 1998. "*Risk Analysis and Management for Projects (RAMP).*" Thomas Telford, London, UK. ISBN 0-7277-2697-8.

Institute of Risk Management (IRM). 2002. "*A Risk Management Standard.*" AIRMIC/ALARM/IRM, London, UK.

Central Computer and Telecommunications Agency (CCTA). 1999. "*Managing Successful Programmes.*" The Stationery Office, London, UK. ISBN 0-11-330026-6.

UK MoD 2002. Ministry of Defence Acquisition Management System, Risk Management Guidance (updated September 2002) www.ams.mod.uk/ams/content/risk/docs/INDEX.HTM.

Norges Standardiseringsforbund (NSF). 1991. Norsk Standard NS5814:1991 "*Krav til risikoanalyser.*"

Office of Government Commerce. 2002. "*Management of Risk—Guidance for Practitioners.*" The Stationery Office, London, UK. ISBN 0-1133-0909-0.

Project Management Institute. 2000. "*A Guide to the Project Management Body of Knowledge (PMBoK).*" 2000 edition. Project Management Institute, Philadelphia, US. ISBN 1-880410-25-7 (CD-ROM).

Simon PW, Hillson DA, Newland KE, eds., 1997. "*Project Risk Analysis and Management (PRAM) Guide.*" APM Group, High Wycombe, Bucks UK. ISBN 0-9531590-0-0.

Standards Australia/Standards New Zealand. 1999. Australian/New Zealand Standard AS/NZS 4360:1999 "*Risk management.*" Standards Australia, Homebush NSW 2140, Australia, and Standards New Zealand, Wellington 6001, New Zealand. ISBN 0-7337-2647-X.

US DoD DSMC 2000. "*Risk Management Guide for DoD Acquisition.*" 3rd ed., January 2000. US Department of Defense, Defense Acquisition University, Defense Systems Management College. DSMC Press, Fort Belvoir, VA 22060-5565, US.

fines risk as "the likelihood of an event, hazard, threat or situation occurring and its undesirable consequences; a potential problem."

Neutral Definitions

More recently other professional bodies have taken a neutral view of risk, such as the UK Association for Project Management (APM), whose 1997 *Project Risk Analysis and Management (PRAM) Guide* defines risk as "an uncertain event or set of circumstances which, should it occur, will have an effect on achievement of the project's objectives." The nature of the impact is undefined, so implicitly this could include both negative and positive effects.

The British Standards Institute echoes this general view in BS6079-3:2000 *Guide to the Management of Business Related Project Risk*, which says that risk is "uncertainty . . . that can affect the prospects of achieving business or project goals."

Another British Standards Institute guide PD 6668:2000 *Managing Risk for Corporate Governance* defines risk in a very open manner, as "chance of something happening that will have an impact upon objectives."

The joint Australian/New Zealand AS/NZS 4360:1999 *Risk Management* standard also has an open definition that could implicitly encompass both opportunities and threats, with risk defined as "the chance of something happening that will have an impact upon objectives."

Finally, British Standard BS IEC 62198:2001 *Project Risk Management Application Guidelines* also defines risk as "combination of the probability of an event occurring and its consequences for project objectives," without specifying whether these consequences are positive or negative.

Broad Definitions

Other recent guidelines explicitly introduce the concept of including upside effects in the definition of risk. For example the British Standard BS ISO 10006:1997 *Quality Management—Guidelines to Quality in Project Management* defines the purpose of risk management as "to minimise impact of potential negative events and to take full advantage of opportunities for improvement," and states that "in this International Standard, the term risk covers both aspects."

The 1997 *Risk Analysis and Management for Projects (RAMP) Guide* from the UK Institution of Civil Engineers states that risk is "a threat (or opportunity) which could affect adversely (or favourably) achievement of the objectives," and this definition is retained in the 2002 update.

Similar terms are found in British Standards BS6079-1:2002 *Guide to Project Management* and BS6079-2:2000 *Project Management Vocabulary*, which define risk as "combination of the probability or frequency of occurrence of a defined threat or opportunity and the magnitude of the consequences of the occurrence."

Chapter 11 of the 2000 edition of the *Guide to the Project Management Body of Knowledge (PMBoK)* produced by the Project Management Institute (PMI) covers project risk management. This states that "Project risk is an uncertain event or condition that, if it occurs, has a positive or negative effect on a project objective . . . Project risk includes both threats to the project's objectives and opportunities to improve on those objectives . . . Project risk management includes maximising the probability and consequences of positive events and minimising the probability and consequences of adverse events to project objectives."

The British Standard guide BSI PD ISO/IEC Guide 73:2002 *Risk Management—Vocabulary—Guidelines for Use in Standards* defines risk as "combination of the probability of an event and its consequence," noting that "consequences can range from positive to negative," and recognizing that "increasingly organisations utilise risk management processes in order to optimise the management of potential opportunities." This definition is also quoted explicitly in *A Risk Management Standard*, published by the UK Institute of Risk Management (IRM) in 2002.

In 2002 the UK Office of Government Commerce launched its *Management of Risk* methodology, which defines risk as "uncertainty of outcome, whether positive opportunity or negative threat."

Also in 2002 the UK Ministry of Defence (MoD) updated its *Risk Management Guidance* website to define risk as "a significant uncertain occurrence . . . defined by the combination of the probability of an event occurring and its consequences on objectives," noting that "the term 'risk' is generally used to embrace the possibility of both negative and/or positive consequences."

TRENDS IN STANDARDIZATION

The first conclusion from this review of national and international risk management standards is that there appears to be no clear consensus on a single accepted definition of risk. The review includes twenty-two standards, of which eight use a negative definition of risk, five take a neutral stance, and nine use a broad definition explicitly including both threat and opportunity.

This of course is the root of the problem, since different standards take different positions. It has rightly been said that "The great thing about standards is that there are so many to choose from!" As a result, risk management professionals engaged in the definition debate are able to refer to one or more standards that support their own preferred position, claiming the authority of their particular chosen standards.

However, it is interesting to note the dates of the various standards in each section reviewed above and in Table 1, as illustrated in Figure 1. It appears that earlier publications tend to use a negative definition of risk, with the more recent ones preferring a neutral or broader approach. Before 1997, all the published risk management standards included in this review used an exclusively negative definition of risk, with the term being synonymous with danger, hazard, loss, etc. From 1997 onward, standards publications started to appear using either a neutral risk definition (where the type of impact is undefined and can therefore implicitly include both negative threats and positive opportunities), or a broad definition explicitly including both types of impact. With the exception of IEEE 1540:2001, none of the standards included in this review that were published since 2000 take the "risk equates to threat" position.

This may suggest a trend away from a threat-only perspective on risk, toward a view that either implicitly allows upside risk or explicitly states that risk covers both threats and opportunities. Clearly future developments must be monitored to determine whether this apparent trend is genuine, or whether those holding the negative position will stage a comeback and resist any broadening of the definition of risk.

A TYPICAL RISK MANAGEMENT PROCESS

It is clear that in principle the term "risk" can be applied equally to opportunities as well as threats, and this is being recognized by the more recently published risk management standards and guidelines. However, in practice the application of this is less readily evident, as the majority of risk management processes remain focused around addressing possible adverse events or circumstances.

For organizations following one of the standards that define risk in wholly negative terms, this is of course not surprising, since this is the only approach allowed by such a definition. Where the guidance being followed

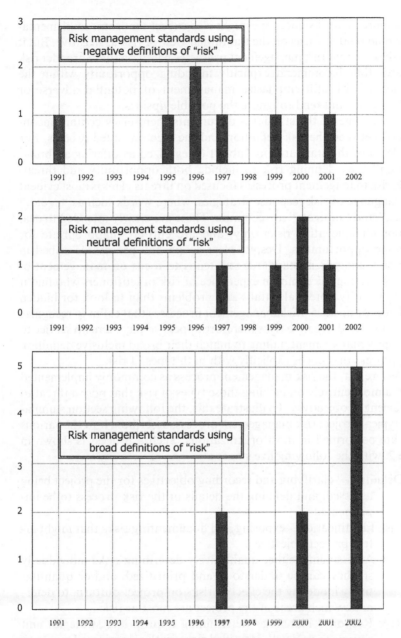

Figure 1 Published risk management standards using different definitions.

takes a more neutral stance, one might also expect that some implementations would tend to focus on the traditional view of risk as negative. But it is also the case that risk management processes based on those broader risk standards that recommend explicitly including opportunity within the definition of risk still emphasize management of potential adverse or harmful events and tend to ignore the possible upsides.

Even some of the standards documents themselves contain inconsistencies between the risk definition and the recommended process. For example all of the standards in Table 1 categorized as offering a "broad definition" (i.e., defining risk to include both opportunity and threat) include risk management processes focused on threats. This is most evident in the language of risk response strategies, where words such as "avoid," "reduce," and "mitigate" are common. These are clearly intended to deal with potential negative risks only, and are entirely inappropriate for addressing opportunities. Despite clear scope, the process described in most risk management standards still tends to focus on management of threats, reflecting the common experience of risk practitioners who find it easier to identify potential pitfalls and problems than to look for hidden advantages or upsides. Risk management processes that claim to recognize both positive and negative risks therefore appear to pay mere lip service to opportunity management, failing to match their broad inclusive definition with a process that copes explicitly with both types of risk.

As a result, the risk management process as commonly implemented focuses almost entirely on tackling those types of risk that pose a threat to achievement of objectives. To illustrate this, the following section summarizes a typical project risk management process, describing how the various stages are performed in most organizations. Such a process is shown in Figure 2, with the following five stages :

Definition—clarifying and recording objectives for the project being assessed, and defining the details of the risk process to be implemented

Risk Identification—exposing and documenting risks that might affect project objectives

Risk Assessment—either qualitatively describing risks individually so they can be understood and prioritized, and/or quantitatively modeling the effect of risks on project outturn, to determine which areas of the project are most at risk

Risk Response Planning—determining appropriate strategies and actions to deal with identified risks, with a nominated owner to address each risk

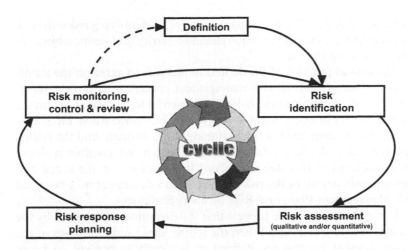

Figure 2 Typical risk management process.

Risk Monitoring, Control, and Review—implementing agreed
responses and checking their effectiveness, reporting to stakeholders, and updating the risk assessment at regular intervals

The first phase is an initiation step, to be completed before the proper risk
process is started. The remaining phases are cyclical, repeated throughout
the life of the project, to ensure that the assessment of risk exposure remains current, and that risk management actions are appropriate and effective.

Definition

The initial phase of the typical project risk management process has two
purposes. The first is to define the objectives for the project being assessed
for risk exposure. This is required because risks cannot be identified in a
vacuum. It is necessary for all stakeholders to understand and agree on the
objectives at stake, prior to initiating the risk process. Although this
sounds obvious, project teams often start trying to identify risks without
a clear understanding of what the project is required to achieve. Equally
often, different stakeholders have conflicting views on the scope or priority
of objectives, and this must be resolved before the risk process can
commence. Objectives should be documented and prioritized in the project
charter or business case, but projects are sometimes required to start work

before this is finalized. To avoid nugatory work in identifying risks, the risk process therefore includes the requirement to clarify and define objectives before proceeding further.

The second purpose of the definition phase is to agree on the scope, aims, and objectives of the risk management process with project stakeholders, as well as defining how risk management is to be carried out on the particular project in operational terms (who, what, how, when, etc.). This phase should be completed at the beginning of the process, and the results communicated to all project stakeholders. Often a process definition document is produced at this stage to record the decisions on the scope and implementation details of the risk process; such a document may be called a Risk Management Plan, or a Risk Strategy Statement.

It is at this point in the process that the definition of risk used by the organization is first relevant. When the scope, aims, and objectives of the risk management process are defined, it is clearly important to know whether this includes only threats, or whether the risk process is expected to address opportunities as well. In most cases the implicit assumption is that the only uncertainties to be covered by the risk process are those that might have a harmful effect on achievement of project objectives. From this starting point, it is therefore inevitable that the remainder of the risk process will ignore opportunities, and only address threats.

Risk Identification

Having defined the risk process to be used for a particular project, the next step is to expose and record all foreseeable risks to project objectives. Risk identification requires forward and creative thinking by project stakeholders. A range of techniques are typically used in this stage, with the most popular being risk workshops or brainstorm sessions involving the project team and other stakeholders. These may be supplemented by use of checklists, prompt lists, interviews, questionnaires, etc.

Risk identification techniques include both group and individual approaches, and both creative and historical methods. Whichever techniques are used, however, the tendency in most projects is to concentrate on identifying potential threats to project objectives. The underlying thought process is focused on what could go wrong, what could prevent the project from making progress or reaching its goals. This is partly due to historical precedent, where most previously identified risks have been negative, and partly due to human nature, which tends to find it easier to criticize or find fault.

As a result, the typical risk identification phase results in a list of uncertain events or circumstances that would have an adverse or unwelcome effect on the project if they were to occur. This focus on threats in the risk identification phase sets the tone for the remainder of the risk process, which continues down the same track, only addressing potential downside. Possible upside risks or opportunities are usually ignored in the risk process, either because the project team are unaware of them, or more rarely because there is a separate opportunity management process to deal with these.

Risk Assessment

The third phase of the typical risk process is risk assessment, which aims to establish the overall level of risk exposure of the project and prioritize identified risks in order of importance.

Risk assessment may be conducted qualitatively or quantitatively, to prioritize risks for management action and to focus management attention on areas most at risk.

Qualitative Assessment

Qualitative risk assessment aims to describe and document each risk in sufficient detail to enable it to be understood and managed. Various characteristics of the risk are determined or estimated, including its cause or trigger conditions, the probability of its occurrence, potential impact on project objectives if it were to occur, areas of the project that might be affected, impact window when the risk might affect the project, horizon when the risk might be no longer able to occur, action window when it is possible to influence the risk, and owner to be responsible for managing the risk and implementing agreed actions. These characteristics are usually recorded in a Risk Register or Risk Log, often held in a computerized database to enable effective data management and reporting. Different techniques are used to evaluate risks based on the qualitative data, usually including prioritization based on probability and impact (perhaps using a two-dimensional matrix or Probability-Impact Grid to present the results). Given that only negative risks (threats) have been exposed during the risk identification phase, the P-I Grid also focuses on prioritizing threats to the project, with the most significant being those with the highest probability of occurring and the most significant negative impact on objectives. The output from this qualitative risk assessment is a prioritized list of risks, allowing management attention to be focused on the most significant risks that could affect the project.

Quantitative Assessment

Quantitative risk assessment, also called quantitative risk analysis, analyzes the combined effect of risks on project outturn using statistical modeling techniques, supported by an appropriate software tool. The most common approach is to use Monte Carlo simulation, based on the project plan, and reflecting the influence of identified risks on planned activities. Risks from the Risk Register are mapped into the project plan, and planned durations or costs are modified to take account of the resulting uncertainty. Multiple iterations of the risk model are then undertaken, randomly sampling from the input ranges of time and/or cost, and allowing mapped risks to affect the project in proportion to their estimated probabilities and impacts. The simulation produces a range of possible project outturns, showing what might happen to the project if risks did or did not occur, and allowing the feasibility of achieving project objectives to be tested. Many projects find that quantitative risk analysis produces results showing a very low chance of achieving targets, predicting overruns in time and/or cost. This is perhaps inevitable given that the risk process is usually focused exclusively on threats, since the only type of effect being modeled in the simulation is negative. As a result, quantitative risk analysis is a one-way street, since the only influence mapped threats can have is to take the project away from its objectives. The analysis is used to expose those areas of the project most at risk, and detailed examination of the data can reveal which activities and risks require most attention to reduce risk exposure.

Risk Response Planning

Once risks have been identified and their significance has been assessed, the next phase of the typical risk process seeks to formulate realistic and effective responses. Responses must be appropriate, affordable, and achievable, taking the significance of each risk into account. Thus severe risks must be the subject of aggressive and urgent attention, whereas less important risks can be treated with less urgency.

A strategic approach to risk response planning is adopted in many risk management processes, with a set of high-level strategies identified. The aim is to select the most appropriate strategy for each risk, depending on its nature, severity, and manageability, and then to design specific actions to implement the chosen strategy. This avoids a scatter-gun random approach where a range of actions are taken without strategic direction, and which may therefore be counterproductive.

Typical response strategies available during the risk response planning phase include avoidance or elimination, transfer, mitigation or reduction, and acceptance.

Avoidance or Elimination

Avoid/eliminate responses seek to remove a particular risk or prevent it from affecting the project. There two types of action under this heading. The first is usually targeted at risk trigger conditions, aiming to remove the cause of the risk, thus making it impossible for the risk to occur. A second type of avoidance response involves executing the project in a different way while still aiming to achieve the same objectives, but without exposing the project to the original risk. Not all risks can be avoided or eliminated, and for others this approach might be too expensive or time-consuming, but this should be the first strategy considered for each risk.

Transfer

Transferring a risk involves finding another party who is willing to take responsibility for its management, and who will bear the liability of the risk should it occur. A range of financial instruments exist to deal with financial risk in this way, and different contractual arrangements can also be made in order to pass risk between the contracting parties. The aim here is to ensure that the risk is owned and managed by the party best able to deal with it effectively. Risk transfer usually involves payment of a premium, and the cost-effectiveness of this must be considered when deciding whether to adopt a transfer strategy.

Mitigation or Reduction

Mitigation or reduction responses aim to modify the size of the risk, by tackling its probability of occurrence and/or its severity of impact. Making a risk less likely or less severe reduces the overall risk exposure of the project. Preventive actions can be designed to reduce the likelihood of a risk occurring, or steps can be taken in advance to protect the project against the effect of a risk should it occur. It is important with this type of response to consider the "risk-effectiveness" of the proposed action, to ensure that cost of achieving the expected reduction in risk probability and/ or impact is not too high. For example, a planned response that would reduce the impact of a cost risk from $10,000 to $1000 but that cost $5000 to implement might be judged less attractive than a second response that would only reduce impact from $10,000 to $5000 but cost just $1000.

Acceptance

Accepting a risk involves either actively making plans for actions to be taken if the risk occurs (i.e., contingency), or passively doing nothing where that is considered appropriate or where no other cost-effective or feasible option exists. Such risks should be built into the project baseline, together with the appropriate level of contingency. They should also be actively monitored to ensure that their severity does not rise to a point where acceptance becomes an inappropriate response.

After these responses have been developed and agreed on, an owner needs to be assigned to each risk response, to take responsibility for implementation of the required action. Responses should also be incorporated into project plans and treated as normal project activities, with owners, budgets, target dates, and resources. There must be a commitment to implement agreed actions as part of the project, and including them in the project plan is the best way to achieve this.

Clearly the various types of risk response are driven by the nature of the risks to be tackled. Since the typical risk management process concentrates exclusively or mainly on threats, it is not surprising that the strategies available in the response planning phase all aim to deal with negative risks only. With the exception of "acceptance," the other strategies (avoidance/elimination, transfer, mitigation/reduction) are not relevant to dealing with opportunities.

Risk Monitoring, Control, and Review

The purpose of the final phase of the iterative risk process is to ensure that agreed risk responses are implemented effectively, to communicate risk status to project stakeholders, and to maintain a current assessment of risk exposure.

This phase involves implementation of agreed responses, to affect risk exposure. Where risk responses have been included in the project plan, they can be reported on and monitored as part of the routine project monitoring process. This should, however, include assessing whether the expected change in risk exposure has been achieved by the responses, or whether additional actions are required, perhaps adopting a different strategy to deal with the risk. Regular risk reviews should be held during the lifetime of the project, to examine the status of previously identified risks, assess the status and effectiveness of agreed risk responses, determine whether new responses are required, and identify new risks.

It is also important to report current risk status to project stakeholders in an appropriate format, to enable appropriate decisions on project strategy to be made in the light of the risk exposure. This may include high-level statements to senior management outlining key risks and current actions, detailed reports to clients and customers to make them aware of the current risk status of the project, and specific targeted information for project team members about risks for which they are responsible.

As with other phases of the typical risk management process, the focus on threats throughout the process means that risk reporting tends to be perceived as "bad news," warning project stakeholders of a range of potential difficulties, some of which will not occur. As a result, the outputs of the risk process are often unwelcome, and may even be discounted or ignored. It is hard to maintain interest in a process that only tells you how bad things might be, even if risk reports also include recommended actions.

THE CURRENT DEBATE AND A PROPOSED SOLUTION

We have seen that risk management practitioners and professionals are engaged in a debate about the definition of risk, discussing whether it should include upside opportunities, or whether it should be confined to equate merely to threats (see the Appendix for a review of this debate). A parallel breadth of definitions is used in official published risk management standards and guidelines (Table 1), reflecting the debate among practitioners. As a result there is no single globally accepted definition of risk, and consequently there is no single "best-practice" approach to implementing risk management in practice.

However, although the standards documentation suggests a possible move toward recognition of a broadened definition of risk to include opportunity, the actual application of risk management seems to be limited to dealing exclusively with threats. This is driven to some extent by inconsistencies in some of the standards documents that espouse a broad definition of risk, but at the same time promote a process where opportunities are not explicitly addressed. It is also a result of common custom and practice, which is not used to including upside thinking. (See Chapter 9 for a more detailed discussion of this issue.)

Clearly this needs to be resolved. One solution is to make the definitions consistent with current practice, and to revert to an approach whereby risk management deals with threats to project objectives, and opportunities are handled separately by a different process. An alternative solution is to modify practice to match the most recent definitions, accepting that thinking has progressed to include both threats and opportunities within the definition of risk, and risk management processes must therefore address both types of uncertainty. A third solution is to do nothing, allowing the current variety of definition and practice to coexist, with no consensus on "best practice."

Those who view risk management as a profession might be uncomfortable with the last option, feeling that to accept the wide diversity of definitions is inconsistent with professionalism. One characteristic of a profession is the presence of a "body of knowledge" that is accepted and followed by all professionals. Consensus on the most basic definitions is surely a prerequisite for such a body of knowledge, as is agreement on the scope and objectives of the core process. Where these are missing, as is currently the case for risk management, it would be hard to argue that the discipline merits being described as a profession.

If the "do nothing" option is undesirable, then we are left with two alternatives: revise the definitions to match current common practice, or modify the practice to bring it into line with current thinking. It appears that trends in standards documentation to include opportunities within the definition of risk are reflecting the thinking of an increasing number of risk management professionals and practitioners, and therefore that reverting to a threat-only definition and process would represent a backward step, which should be avoided in any discipline committed to development and progress.

There is, however, a problem with taking the option that accepts the broadening definition of risk and attempts to similarly widen the application of the risk management process. This arises from the undoubted focus of most implementations of risk management on dealing solely with threats. The typical risk management process (as described above) is only concerned with addressing uncertainties with possible negative effects, even where the possibility of upside risk is theoretically accepted.

Addressing this problem will require changes to the standard approach to risk management. While the five-phase process of Definition, Risk Identification, Risk Assessment, Risk Response Planning, and Risk

Monitoring/Control/Review is still applicable, each phase needs to be considered to ensure that it is capable of handling opportunities alongside threats. A common approach would have many benefits, including lack of additional process overhead, building on accepted techniques, using the same resources and time, improving efficiency with double "bangs per buck," extending the existing "uncertainty" mind-set, and ensuring that opportunities are managed proactively.

However, if the existing risk management process is to be extended to allow opportunities to be managed alongside threats, some changes will be required. Part II of this book suggests where such process modifications should be focused, followed in Part III by discussion of a number of implementation issues that are essential to take advantage of a broadened process.

REFERENCES

Australian/New Zealand Standard AS/NZS 4360:1999. (1999). *Risk Management*. Published jointly by Standards Australia, Homebush NSW 2140, Australia, and Standards New Zealand, Wellington 6001, New Zealand.

Boothroyd, C. E., Emmett, J. (1996). *Risk Management—A Practical Guide for Construction Professionals*. London:Witherby.

British Standard BS6079-1:2002. (2002). *Project Management—Part 1: Guide to Project Management*. London: British Standards Institute.

British Standard BS6079-2:2000. (2000). *Project Management—Part 2: Vocabulary*. London: British Standards Institute.

British Standard BS6079-3:2000. (2000). *Project Management—Part 3: Guide to the Management of Business-Related Project Risk*. London: British Standards Institute.

British Standard BS8444-3:1996. (1996). (IEC 300-3-9:1995). *Risk Management: Part 3—Guide to Risk Analysis of Technological Systems*. London: British Standards Institute.

BS IEC 62198:2001. (2001). *Project Risk Management—Application Guidelines*. London: British Standards Institute.

BSI PD 6668:2000. (2000). *Managing Risk for Corporate Governance*. London: British Standards Institute.

BSI PD ISO/IEC Guide 73:2002. (2002). *Risk Management—Vocabulary—Guidelines for Use in Standards*. London: British Standards Institute.

CCTA. (1993). *Introduction to the Management of Risk*. London: HMSO.

CCTA. (1994). *Management of Project Risk*. London: HMSO.

CCTA. (1995). *Management of Programme Risk*. London: HMSO.

Chapman, C. B. (1997). Project risk analysis and management—PRAM, the generic process. *Int. J. Project Management* 15(5):273–281.

Collins. (1979). *Collins Dictionary of the English Language*. Glasgow, UK: William Collins Sons, Co Ltd.

Defense Systems Management College. (1989). *Risk Management—Concepts and Guidance*. Fort Belvoir, VA: DSMC Press.

Dorofee, A. J., et al. (1996). *Continuous Risk Management Guidebook*. Pittsburgh, PA: Software Engineering Institute, Carnegie Mellon University.

Engineering Council. (1995). *Guidelines on Risk Issues*. London: Engineering Council.

Godfrey, P. S. (1995). *Control of Risk: A Guide to the Systematic Management of Risk from Construction*. (CIRIA under reference FR/CP/32). London: CIRIA.

HM Treasury. (1993). Managing risk, contingency for defence works projects. CUP Guidance Note No. 41.

HM Treasury. (1994). Risk management guidance note, June 1994.

Hulett, D. T., Hillson, D. A., Kohl, R. (2002). Defining Risk: A Debate. *Cutter IT J.* 15(2):4–10.

Institution of Civil Engineers (ICE) and Faculty, Institute of Actuaries. (1997). *Risk Analysis, Management for Projects (RAMP)*. London, UK: Thomas Telford.

Kähkönen, K., Arrto, K. A. (2000). Balancing project risks and opportunities. In: *Proceedings of the 31st Annual Project Management Institute Seminars, Symposium* (PMI 2000), presented in Houston, TX, 7–16 September 2000.

Moore, A. Fearon, A., Alcock, M. (2001). Implementation of opportunity and risk management in BAE Systems Astute Class Limited—a case study. In: *Proceedings of the 4th European Project Management Conference* (PMI Europe 2001), presented in London, UK, 6–7 June 2001.

National Standard of Canada CAN/CSA-Q850-97. (1997). *Risk Management: Guideline for Decision-makers*. Ontario, Canada: Canadian Standards Association.

Norges Standardiseringsforbund (NSF). (1991). Norsk Standard NS5814:1991 Krav til risikoanalyser.

Project Management Institute. (2000). *A Guide to the Project Management Body of Knowledge (PMBoK®)*. 2000 ed. Philadelphia: Project Management Institute.

Simon, P. W., Hillson, D. A., Newland, K. E., eds. (1997). *Project Risk Analysis, Management (PRAM) Guide*. High Wycombe, Bucks, UK: APM Group.

UK Association for Project Management. (2000). *Project Management Body of Knowledge*. 4th ed. High Wycombe, Bucks, UK: APM.

UK Office of Government Commerce (OGC). (2002). *Management of Risk—Guidance for Practitioners*. London: Stationery Office.

US Department of Defense. (January 2000). *Risk Management Guide for DoD Acquisition*. 3rd ed. Defense Acquisition University, Defense Systems Management College. Fort Belvoir, VA: DSMC Press.

Ward, S. C. (1999). Requirements for an effective project risk management process. *Project Management J.* 30(3):37–43.

II
THE OPPORTUNITY
MANAGEMENT PROCESS

3
Definition

The second part of this book works through the typical project risk management process as outlined in Chapter 2, and indicates where modifications to the process are required to allow opportunities to be managed proactively and effectively alongside threats. The aim is to produce a broad common risk management process that is capable of handling both threats and opportunities together, i.e., dealing with both types of risk (see Chapter 1) in the same way.

The first phase of any risk management process is Definition, acting as an initiation step prior to embarking on the iterative phases of Risk Identification, Risk Assessment, Risk Response Planning, and Risk Monitoring/Control/Review. This chapter describes the typical Definition phase, focusing on those aspects that may need to be changed to include opportunities within the scope of the risk process.

SETTING EXPECTATIONS

Early risk management processes started with risk identification as the first phase, reasoning that risks could not be managed if they were not first identified. But we have already emphasized the importance of clearly defining what is meant by the term "risk," to avoid confusion and wasted effort. Defining risk as "Any uncertainty that, if it occurs, would affect one or more project objectives" brings together the two dimensions of uncertainty and effect on objectives. But this definition reveals a prerequisite that must be addressed before risks can be identified. If a risk is defined only in relation to project objectives, then those objectives must themselves be defined before any risks can be identified.

As a result, best-practice risk processes now start with a Definition phase, to be completed before risk identification. This ensures that project objectives are agreed on, understood, and communicated within the project stakeholder community. Another purpose fulfilled by the initial Definition phase is to decide the level of detail required for the risk process, driven by the riskiness and strategic importance of the project. Some projects may only require a simple risk process, whereas others will need more in-depth risk management; the Definition phase should describe the level of detail of the process to be implemented.

The purpose of the Definition phase is therefore to ensure that all project stakeholders share a common view of the project's objectives, are aware of the aims and objectives of the risk management process, and understand the agreed approach to managing risk on this particular project. The key outputs from this phase are not documents or reports, but awareness, consensus, and understanding. As a result, it is important that this phase is unambiguous and clear, involving the key project participants in the Definition process, and not moving into the iterative part of the risk process until Definition is complete.

There are obvious dangers in proceeding with risk management without agreement or understanding over project objectives, or lack of consensus concerning the scope and objectives of the risk process. Different stakeholders might take contradictory or conflicting positions during the risk process if they have inconsistent perspectives on the aims and objectives of risk management for their project. For example, lack of agreed risk thresholds can lead to unclear risk strategies as one stakeholder may feel unable to accept risks with which another is entirely comfortable. Differences of opinion over the scope of the risk process can result in some stakeholders seeking to include aspects of the project within the risk process that others believe should be excluded, or attempting to deal with types of risk that other stakeholders are not addressing.

This need for shared understanding at the commencement of the risk process is met by the Definition phase, by documenting key aspects of risk management as they are to be applied to a specific project. This is usually achieved by production of a document known as the Risk Management Plan or Risk Strategy Statement (see below), recording important elements to define the implementation of risk management for a particular project. Issue of this document allows the expectations of stakeholders to be brought into alignment, and forms the basis for future management of those expectations during the risk process.

At first thought it may seem as if no changes are required to the standard Definition phase to account for opportunities within the risk process. However, the main goal of this phase is to ensure that all stakeholders in the risk process share a common perspective concerning the scope and objectives of risk management for the project under consideration. Part of that consensus is a shared understanding of the nature of risk itself, and what risks are to be included in the scope of the risk process for this particular project. In the absence of any explicit statement to the contrary, most project stakeholders are likely to assume the "traditional" view of risk, namely that it represents only adverse, harmful, negative, or unwelcome potential events. If this perception is allowed to persist unchallenged, the tone will be set for the remainder of the risk management process to focus only on threats.

As a result, if the risk management process is to be extended to cope with opportunities as well as threats, it is essential that the Definition phase makes this explicit to set appropriate expectations among the stakeholder community. By stating clearly that the organisation uses a definition of risk that includes both opportunities and threats, the Definition phase will ensure that the risk process is initiated with this in mind, thus setting the scene for proactive management of both upside and downside risks to the project.

DEFINING OBJECTIVES AND SCOPE

The purpose of the Definition phase is simple, and matches the standard approach to project management itself. In the same way that a project cannot commence without a clear statement of objectives, so the risk process also requires an initiation phase to determine and communicate its boundaries and purpose. However, contrary to the theoretical best-practice position, there are many projects where the project team do indeed start work without a clear understanding of their purpose. Despite the official or formal requirement for a project charter or business case as a prerequisite to commencement, the pressure to start the project often proves irresistible and the project team launch into the detailed tasks. One of the most common reasons for project failure has been found to be unclear objectives or lack of clear requirements. The project charter is intended to preclude this by insisting that objectives and requirements are clearly defined before the project starts work, but unfortunately reality does not always match theory.

Since many of the more recent project risk management processes were produced by people familiar with the disciplines and practice of project management itself, it is perhaps not surprising that they transcribed the requirement for an initiation phase into risk management. The Definition phase (under a variety of names) has therefore become a standard feature of most risk processes, following the example set by project management. It serves several purposes in laying the groundwork for initiation of the risk process, as outlined in the following sections.

Project Objectives

The first purpose of Definition is to define the objectives for the particular project. All stakeholders must understand and agree on the objectives at stake, prior to initiating the risk process. Different stakeholders may have conflicting views that must be resolved. This should be possible by simple reference to the project charter or business case in which project objectives should be documented and prioritized, but projects are sometimes required to start work before this is finalized. To address any possible shortfall in the project initiation process, the risk process therefore includes the requirement to clarify and define project objectives before proceeding further. If this has already been done elsewhere in the project process, a cross-reference or summary will suffice; otherwise project objectives should be explicitly stated at this point in the risk process.

Risk Process Scope

The Definition step in the risk process also seeks to document the scope, aims, and objectives of the risk management process, and communicate this to project stakeholders so that they can agree together. This should include a description of how risk management is to be carried out on the particular project in operational terms (who, what, how, when, etc.).

It is important to define the scope of the risk process before it commences so that subsequent steps are properly focused. A typical project will be undertaken in a complex environment involving a range of dependencies and stakeholders, as illustrated in Figure 1. However, the risk process

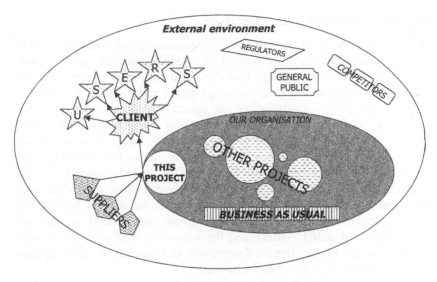

Figure 1   Elements of project environment to consider for risk process scope.

may be limited in scope to cover a subset of the entire environment. In the most simple case, the scope of risk management might only cover internal responsibilities of the project team, excluding all other aspects of the project environment. Or the organization may choose to include client risks or supplier risks within the scope of the project's risk process. Risks affecting other projects or business as usual might also be taken into scope, or wider corporate and strategic issues could be included.

Another important element of scope definition is to determine what types of risks will be included in the risk process. We have seen in Chapter 1 that projects are subject to a wide variety of sources of risk (see, for example, Table 1, in that chapter), and the organization may decide that these are not all to be encompassed within the risk process for a particular project. For example, it may be deemed appropriate on a project to limit risk management effort to dealing with technical and management risks and to exclude risks originating externally to the project.

Without clearly defining the scope of the risk management process, risk identification will be confused and unfocused, it will be difficult to determine applicable assessment criteria, development of responses and allocation of owners will be hindered, and the overall effectiveness of the risk process will be compromised.

Risk Process Aims and Objectives

Aims and objectives for the risk process are also required, since these might also vary between different projects. For example, a project may be undertaken with the clear knowledge that it is more risky than average, but the organization may be prepared to accept higher risk to gain the potential benefits. In this case the aim of the risk process might be to allow a higher degree of risk to be taken but to ensure that it is managed effectively. Another project might be more routine, where the risk process would aim to ensure no escalation of risk above acceptable limits. For a third project, the aim of the risk process might be to manage the interface between the organization and its supply chain to minimize exposure to possible market variations. Having defined the specific goals of the risk process for each particular project, stakeholders and project team members alike will know what is expected of them, and the success or otherwise of risk management on the project can be determined and monitored, allowing corrective action to be taken where required.

Risk Process Characteristics

The last main purpose of the Definition phase is to define and describe the characteristics of the risk process to be implemented for the particular project in question. This includes outlining the methodology, approach, tools, and techniques to be followed, and identifying roles and responsibilities for risk management among project stakeholders. Reporting standards, formats, and frequencies should be stated, and review and update cycles defined. This is important because risk management is not a "one-size-fits-all" discipline. The level of risk management required will vary from one project to another. Many factors drive the required level of risk management for a particular project including:

> The riskiness of the project
> The strategic importance of the project
> The risk appetite of the organization and its management team
> The relative risk exposure of other current projects within the organization (given the desire to create a risk-balanced portfolio)
> The maturity and capability of the organization in managing risk and projects
> The degree of resources available for risk management

Some projects will require a very detailed and in-depth risk process, with multiple techniques and expert resources, frequent reviews, and a wide range of inputs, specialized tools, and leading-edge techniques. For others a more simplified approach will be appropriate, perhaps conducting risk assessments as part of the routine project management tasks, with no special resources or tools. It is a key purpose of the Definition phase to answer these questions, scaling the risk process as appropriate to meet the specific needs of the project. Some organizations may decide to adopt a formal project sizing approach, assessing project characteristics against a set of criteria to determine whether the project is large/medium/small, or some other measure of complexity and/or riskiness. The level of risk management on the project can then be set according to the "project size" determined from this assessment, with typical projects using a standard risk process, large or complex projects having an extended risk process, and a simplified risk process being used for small or simple projects.

DOCUMENTING REQUIREMENTS: THE RISK MANAGEMENT PLAN

The key output from the Definition phase is a document capturing these various aspects discussed above, namely project objectives, risk process scope, risk process aims and objectives, and risk methodology and approach. The document can then be used to communicate the parameters of the risk process to relevant stakeholders, and can also act as a reference during execution of the remainder of the risk process. Many times those members of project teams involved in the risk process have no clear guidance on what they are supposed to be doing. They are therefore unable to determine whether they are succeeding in making the risk process effective, since they have no clear target at which to aim. Documenting the results of the Definition phase provides a statement of how risk management should work for this particular project, allowing process effectiveness and efficiency to be measured.

The document describing the risk process to be implemented for a particular project may be called one of several names, but perhaps the most common is the Risk Management Plan. This is not to be confused with any document listing risks and responses, which might be known as a Risk Register or Risk Response Plan. Instead the Risk Management Plan is a process description document, defining the process to be followed. In this respect it is similar to a Quality Plan (outlining the quality process, but not including the results of audits) or a Resource Management Plan (defining the approach to resourcing rather than presenting actual resource levels during the project).

The level of detail contained in a Risk Management Plan can of course vary according to the requirements of the project or the documentation standards of the organization or client. In some cases a Risk Management Plan may be quite short, summarizing the essential elements on one or two pages, perhaps created by completing sections within a generic template. For other projects, the Risk Management Plan may be required to present considerable detail as a stand-alone document.

Whatever format is used, it is important that the Risk Management Plan describes the risk process unambiguously, stating the requirements for risk management on the particular project so that all project stakeholders know what is expected of them, and so that the organization can determine whether the risk process is succeeding in meeting its objectives.

At the end of this chapter an example of a typical Risk Management Plan for a mythical project is presented, highlighting those areas where specific statements are required to ensure that the risk process covers opportunities adequately alongside threats. This example is fairly detailed, recognizing that for many projects a shorter document will suffice to define the project objectives and risk process characteristics.

GAINING STAKEHOLDER COMMITMENT

This chapter has described the initiation phase for the risk management process, called here the Definition phase. It serves a wide range of purposes in ensuring that the risk process is properly focused and targeted, documenting key aspects of risk management and communicating them to project stakeholders.

However, Definition activities should not be completed simply to produce a document for the project library. The result of a successful Definition phase is awareness and shared understanding of the purpose and scope of the risk process for a particular project, combined with agreement and consensus on how risk management should be implemented in this instance. Once the document has been produced, it should be used by the Project Manager and Project Sponsor to gain and retain commitment from project stakeholders to managing risk effectively on their project. This requires wide circulation of the Risk Management Plan, with follow-up to ensure that it has been read and absorbed, and that there is a determination to act in accordance with the provisions contained in the plan.

This is particularly important when the organisation wishes to include management of opportunities as part of the risk management process. If the Risk Management Plan explicitly describes how opportunities will be addressed as part of the process, it is vital for participants in the risk process to be aware of this and buy into it. This is unlikely to occur if the Risk Management Plan remains on the shelf, unread by those who have to follow its provisions. Where an organization is determined to manage opportunity alongside threat during the risk process, proactive follow-up is required to encourage project stakeholders to take this seriously. The Project Manager and/or Project Sponsor should consider visiting key stakeholders with the Risk Management Plan, to discuss it in person with them, and to emphasize the opportunity aspects contained in it.

There is an uphill battle to be fought here, to overcome the natural tendency to focus exclusively on the negative and to ignore opportunities. Left to their own devices, those involved in the risk management process are likely to think only of threats, even if the Risk Management Plan says something different. Merely writing something down does not guarantee that it will happen. The Risk Management Plan only defines the characteristics of the desired risk process. But just stating it does not ensure that participants in the project will either understand what is wanted, or commit to meeting it. As a stand-alone document, the Risk Management Plan is merely a set of propositions and aspirations. The Project Manager may wish to consider using some simple tools to move beyond this position, for example a facilitated workshop to work through the Risk Management Plan with the relevant parties, exploring the implications of each point, ensuring understanding, and reaching consensus. In particular, proactive steps are required to ensure that those participating in risk management on a project share a common understanding that risk includes both opportunity and threat, and that both require effective management if the project is to maximize its chances of achieving its objectives.

Once this Definition phase is complete, the risk process can continue into the iterative part, starting with Risk Identification. Having set the expectation at the outset that risk management should include dealing with opportunities, the remainder of the cyclical process stands a chance of achieving this aim. The next chapter explores how to ensure that opportunities are identified alongside threats, enabling them to be managed and captured to benefit the project.

REFERENCES

Australian/New Zealand Standard AS/NZS 4360:1999. (1999). *Risk Management*. Published jointly by Standards Australia, Homebush NSW 2140, Australia, and Standards New Zealand, Wellington 6001, New Zealand.

British Standard BS6079-3:2000. (2000). *Project Management—Part 3: Guide to the Management of Business-Related Project Risk*. London: British Standards Institute.

De Bakker, K., Stewart, W. M., Sheremata, P. W. (19–20 June 2002). Risk management planning—how much is good enough? In: *Proceedings of the 5th European Project Management Conference* (PMI Europe 2002), presented in Cannes, France.

Institution of Civil Engineers (ICE) and Faculty, Institute of Actuaries. (1997). *Risk Analysis, Management for Projects (RAMP)*. London: Thomas Telford.

Project Management Institute. (2000). *A Guide to the Project Management Body of Knowledge (PMBoK®)*. (2000 ed). Philadelphia: Project Management Institute.

Project Management Institute. (2001). *Practice Standard for Work Breakdown Structures*. Philadelphia: Project Management Institute.

Simon, P. W., Hillson, D. A., Newland, K. E., eds. (1997). *Project Risk Analysis, Management (PRAM) Guide*. High Wycombe, Bucks, UK: APM Group.

UK Office of Government Commerce (OGC). (2002). *Management of Risk—Guidance for Practitioners*. London: Stationery Office.

SAMPLE RISK MANAGEMENT PLAN FOR THE MEGATRONIC ENHANCEMENT PROJECT

Prepared by: I M Responsible (Project Manager)
Approved by: U Shoraar (Project Sponsor)
Reference: MEP/12–34/imr   Version: 1   Date: 1 April 2004

DOCUMENT PURPOSE

This document is the Risk Management Plan for the Megatronic Enhancement Project, defining the risk management process to be employed throughout the life of this project. The Project Manager is responsible for reviewing and maintaining this Risk Management Plan throughout the project, to ensure that the risk process remains appropriate to deal with the level of risk faced by the project.

PROJECT DESCRIPTION AND OBJECTIVES

The Megatronic product range is the current market leader in combined electronic communications and leisure devices for older children aged 8–11 in North America (United States and Canada). The Megatronic Enhancement Project has been initiated by Projects-R-Us to provide a platform for extending our existing Megatronic product range into the teenage Western European market, to take advantage of recent changes in demographics and spending power. The main objective is to develop enhancements to current products that will appeal to the new market segment, thus allowing the organization to increase market share and produce a new profitable revenue stream.

As a result, the Megatronic Enhancement Project is perceived as a medium-risk project, but it is regarded as very significant for the future strategic positioning of Projects-R-Us.

The scope and objectives for the Megatronic Enhancement Project are described in full in the Project Charter (reference MEP/56–78/imr dated 31 January 2004), and are summarized as follows:

- Perform detailed design studies to implement changes to the Megatronic product range as recommended by recent market research, to facilitate sales into the new sector of Western European teenagers.
- Produce a trial batch of modified products for market testing.
- Identify and cost changes required to current Megatronic production facilities to enable mass manufacture of modified product range, increasing capacity by 150%.
- Project start date is 30 April 2004. Production of trial samples of modified products must be completed by 30 September 2004, to enable marketing campaign and full-scale production leading up to full launch in time for the Christmas 2004 season.

- Project budget has been set at USD $150,000, to cover all project costs, including design studies and production of trial batch, but excluding production line modifications or the marketing campaign.
- Continue to meet all quality, safety, environmental, and regulatory requirements of existing market, as well as meeting the requirements of the new market segment.
- Ensure that new products are favorably positioned relative to current market competitors, particularly in terms of new functionality.
- The Megatronic Enhancement Project covers modification of all current elements in the Megatronic product range, but no new products are to be developed.

Project objectives are prioritized with time as the most important, followed by functionality and quality, then finally cost.

DEFINITIONS

The following definitions apply to the Megatronic Enhancement Project risk management process:

Risk: Any uncertainty that, if it happens, could have a positive or negative effect on one or more project objectives. [Note: This definition of risk explicitly includes uncertainties with positive effects (also called "opportunities") as well as those with negative effects (or "threats").]

Risk Management: The proactive structured management process for ensuring that risks are identified, assessed, and responded to appropriately. [Note: The risk management process is expected to deal equally effectively with both opportunities and threats.]

Opportunity: A risk with a positive impact.

Threat: A risk with a negative impact.

Risk Register: The document listing and describing all identified risks together with responses and their owners. [Note: Opportunities and threats must both be listed in the same Risk Register, but will be clearly distinguished from one another.]

Risk Response: An action to be taken to deal appropriately with an identified risk. [Note: Proactive risk responses are required for both opportunities and threats. Appropriate responses for opportunities include exploiting, sharing, enhancing, or accepting. Threat response options include avoiding, transferring, reducing, or accepting.]

Probability: Estimate of the uncertainty associated with a risk, describing how likely it is to occur, expressed either descriptively (e.g., Very Low/Low/Medium/High/Very High—see Appendix A) or numerically as a percentage (e.g., in the range 1–99%).

Impact: Estimate of the effect that a risk would have if it occurs, measured against one or more project objectives. This is expressed either descriptively (e.g., Very Low/Low/Medium/High/Very High—see Appendix A) or numerically (e.g., days, USD$, etc.) [Note: Impact is defined as negative for threats, for example time delay, increased cost, reduced functionality, etc. When assessing opportunities, impact is defined in positive terms, for example early delivery, cost saving, increased functionality, etc.]

Risk Score: A dimensionless number used to prioritize risks. Risk Scores range between 0 and 1, and are calculated by multiplying a probability value and an impact value (both between 0 and 1). Probability values are defined as 0.1/0.3/0.5/0.7/0.9 for Very Low/Low/Medium/High/Very High, respectively. Impact values are defined as 0.05/0.1/0.2/0.4/0.8 for Very Low/Low/Medium/High/Very High, respectively.

Risk Breakdown Structure (RBS): A hierarchical grouping of sources of risk. The RBS for this project is shown in Appendix B.

AIMS AND OBJECTIVES OF RISK PROCESS

The Megatronic Enhancement Project risk management process aims to manage all foreseeable risks (both opportunities and threats) in a manner that is proactive, effective, and appropriate, in order to maximize the likelihood of the project achieving its objectives, while maintaining risk exposure at an acceptable level.

"Acceptable risk" is defined for the Megatronic Enhancement Project as a balance of opportunities and threats, with no more than five identified threats remaining in the "high risk" category (i.e., Risk Score >0.20) after suitable responses have been implemented, and a total Risk Score for opportunities at least 20% higher than the total Risk Score for threats.

The risk process will aim to engage all project stakeholders appropriately, creating ownership and buy-in to the project itself and also to risk management actions.

Risk-based information will be communicated to project stakeholders in a timely manner at an appropriate level of detail, to enable project strategy to be modified in the light of current risk exposure.

The risk management process will enable project stakeholders to focus attention on those areas of the project most at risk, by identifying the major risks (both opportunities and threats) potentially able to exert the greatest positive or negative influence on achievement of project objectives.

SCOPE OF RISK PROCESS

The risk management process to be implemented on the Megatronic Enhancement Project will deal equally with opportunities and threats, and both types of uncertainty are to be explicitly included within the scope of the process at all times.

The risk management process covers all design, development, and production activities during the lifetime of the project.

All risks internal to the Megatronic Enhancement Project will be included in the scope of the risk management process, and this process will also include risks arising from members of the supply chain and subcontractors. This risk process will not deal with risks outside the immediate scope of the Megatronic Enhancement Project, e.g., corporate risks, or risks affecting other projects or business-as-usual activities. If such risks are identified they will be escalated to senior management. The process also excludes end-user risks (except product functionality risks).

Risks affecting the planned market testing of enhanced products after completion of the Megatronic Enhancement Project may be identified and recorded during this project, but they will not be actively managed as part of this risk management process.

Sources of risks to be considered during the risk management process include technical, management, commercial, and external, as defined in the standard Projects-R-Us Risk Breakdown Structure (see Appendix B). Particular attention will be given to quality, safety, environmental, and regulatory risks.

RISK METHODOLOGY AND APPROACH

The standard Projects-R-Us risk methodology will be used for the Megatronic Enhancement Project, following the five phases outlined in company procedure PRU-RM-01 "Project Risk Management." These are summarized as follows:

> *Definition*—clarifying and recording objectives for the project being assessed, and defining the details of the risk process to be implemented, documenting the results in a Risk Management Plan
>
> *Risk Identification*—exposing and documenting risks that might affect project objectives either positively or negatively
>
> *Risk Assessment*—either qualitatively describing risks individually so they can be understood and prioritized, and/or quantitatively modeling the effect of risks on project outturn, to determine which areas of the project are most at risk
>
> *Risk Response Planning*—determining appropriate strategies and actions to deal with identified risks, with a nominated owner to address each risk
>
> *Risk Monitoring, Control, and Review*—implementing agreed responses and checking their effectiveness, reporting to stakeholders, and updating the risk assessment at regular intervals

For the Megatronic Enhancement Project, risk assessment will be limited to qualitative methods, with no requirement for quantitative risk modeling.

The Definition phase will be completed before the project commences, then the remaining phases of the risk process will be cyclical, repeated regularly throughout the life of the project. The initial cycle will be completed within one month of project start, and updates will be performed monthly thereafter.

RISK TOOLS AND TECHNIQUES

The following tools and techniques will be used to support the risk management process on the Megatronic Enhancement Project:

Definition

Risk Management Plan (this document), issued at project start and reviewed by the Project Manager three-monthly during the project.

Risk Identification

- Risks (both threats and opportunities) will be identified using the following techniques:

  Brainstorming with all members of the project team plus representatives of key suppliers
  Review of the standard risk checklist PRU-RM-04
  Query of lessons learned from Project Knowledge Database
  Ad hoc identification of risks by project team members at any time during the project, raising risks using standard risk capture form PRU-RM-06

- Preliminary Risk Register to record identified risks for further assessment, following the standard format (see PRU-RM-07).

Risk Assessment

- Probability and Impact Assessment for each identified risk, using the project-specific scales defined in Appendix A.
- Dual P-I Grid to prioritize risks for action, using the standard Risk Scoring calculations based on probability (P) and impact (I).
- "Top Ten" Risk List for priority management attention.
- Risk categorization using the standard Risk Breakdown Structure (see Appendix B) to identify patterns of exposure.
- Risk Register update to include assessment data.

Risk Response Planning

- Response Strategy Selection as appropriate for each identified risk, including owner allocation.
- Risk Register update to include response data.

Risk Monitoring, Control, and Review

- Project Review Meeting to review status of key risks.
- Risk Assessment Report to senior management.
- Risk Review Meeting to identify new risks, review progress on existing risks and agreed responses, and assess process effectiveness.

Details of each of these tools and techniques are contained in appendices to the company procedure PRU-RM-01 "Project Risk Management."

ROLES AND RESPONSIBILITIES FOR RISK MANAGEMENT

The responsibilities of key project stakeholders for risk management on the Megatronic Enhancement Project are defined in individual Terms of Reference for each job role, and summarized as follows:

Project Manager

- Overall responsibility for the risk management process, to ensure that foreseeable risks (both threats and opportunities) are identified and managed effectively and proactively to maintain an acceptable level of risk exposure for the project.
- Leading the project team in the risk process, including chairing project reviews and risk review meetings.
- Selecting appropriate tools and techniques for the risk process, and documenting these decisions in the Risk Management Plan.
- Approving risk response plans prior to implementation.
- Applying project contingency funds to deal with identified risks that occur during the project.
- Overseeing risk management by subcontractors and suppliers.
- Reporting current risk status on the project to senior management, with recommendations for appropriate strategic decisions and actions to maintain acceptable risk exposure.
- Highlighting to senior management any identified risks that are outside the scope or control of the project, or that require input or action from outside the project, or where release of "management reserve" funds might be appropriate.

Project Sponsor

- Actively supporting and encouraging the implementation of a formal risk management process on the project.
- Reviewing risk outputs from the project with the Project Manager to ensure process consistency and effectiveness.
- Reviewing risks escalated by the Project Manager that are outside the scope or control of the project or that require input or action from outside the project.
- Taking decisions on project strategy in the light of current risk status, to maintain acceptable risk exposure.
- Ensuring adequate resources are available to the project to respond appropriately to identified risk.
- Releasing "management reserve" funds to the project where justified to deal with exceptional risks.

Project Team Member

- Participating actively in the risk process, proactively identifying and managing risks in their area of responsibility.
- Implementing agreed responses in a timely and professional manner, and reporting status to the Project Manager.
- Providing inputs to the Project Manager for risk reports.

Subcontractor/Supplier

- [Same as Project Team Member]

RISK REVIEWS

Risk exposure on the Megatronic Enhancement Project will be reviewed monthly during the life of the project. Risk Review Meetings will be held at which new risks will be identified and assessed, existing risks will be reviewed, progress on agreed actions will be assessed, and new actions and/or owners will be allocated where required.

The effectiveness of the risk process will be reviewed as part of the Risk Review Meeting at three-monthly intervals during the project, to determine whether changes to the approach, tools, or techniques are required. Where process changes are agreed by the Project Manager, this Risk Management Plan will be updated and reissued to document the revised process.

Progress on key risks will be reviewed at the monthly Project Review Meeting.

RISK REPORTING

A Risk Assessment Report will be issued monthly by the Project Manager to the Project Sponsor after each Risk Review Meeting, with the following content:

> Executive Summary
> Project status and overall risk status
> Key risks and current responses
> Changes since last report
> Risks outside project scope or control
> Risk metrics and trend analysis
> Conclusions and recommendations
> Appendix: Current Risk Register

Project team members, subcontractors, and suppliers will be provided with an extract from the current Risk Register after each monthly Risk Review Meeting, listing those risks and actions for which the individual is responsible.

On completion of the project, a risk section will be provided for the Megatronic Enhancement Project Lessons Learned Report, detailing generic risks (both opportunities and threats) that might affect other similar projects, together with responses that have been found effective in this project. Input will also be provided for the Project Knowledge Database, to capture risk-related lessons learned from this project.

APPENDIX A: DEFINITIONS OF PROBABILITY AND IMPACT

| Scale | Probability | +/−Impact on Project Objectives | | |
		Time	Cost	Performance
VHI	76–95%	>20 days	>$20K	Very significant impact on overall functionality
HI	61–75%	11–20 days	$11K–$20K	Significant impact on overall functionality
MED	41–60%	4–10 days	$6K–$10K	Some impact in key functional areas
LO	26–40%	1–3 days	$1K–$5K	Minor impact on overall functionality
VLO	5–25%	<1 day	<$1K	Minor impact on secondary functions
NIL	<5%	No change	No change	No change in functionality

[Note: When these impact scales are used to assess opportunities, they are to be treated as representing a positive saving in time or cost, or increased functionality. For threats, each impact scale is interpreted negatively, i.e., time delays, increased cost, or reduced functionality.]

APPENDIX B: STANDARD PROJECTS-R-US RISK BREAKDOWN STRUCTURE (RBS)

RBS Level 0	RBS Level 1	RBS Level 2
0. Project risk	1. Technical risk	1.1 Scope definition
		1.2 Requirements definition
		1.3 Technical processes
		1.4 Technology
		1.5 Technical interfaces
		1.6 Technology scaling
		1.7 Performance
		1.8 Reliability, safety, and security
		1.9 Test and acceptance
	2. Management risk	2.1 Project management
		2.2 Organization
		2.3 Resourcing
		2.4 Communication
		2.5 Information
		2.6 HS and E
		2.7 Reputation
	3. Commercial risk	3.1 Contractual T and Cs
		3.2 Financing
		3.3 Liabilities and warranties
		3.4 Payment terms
		3.5 Suspension and termination
		3.6 Internal procurement
		3.7 Subcontracts
		3.8 Client stability
		3.9 Applicable law
		3.10 Partner financial stability
		3.11 Partner relevant experience
	4. External risk	4.1 Legislation
		4.2 Exchange rates
		4.3 Site/facilities
		4.4 Competition
		4.5 Regulatory
		4.6 Political
		4.7 Country
		4.8 Pressure groups
		4.9 Force majeure

4
Identification

Risk cannot exist in a vacuum. We can only speak of a "risk to" something, and in project terms, we are interested only in risks to project objectives. This is captured in our definition of a risk as "Any uncertainty which, if it occurs, would affect one or more project objectives." An essential prerequisite to being able to identify risks is therefore a clear understanding of project objectives. This is delivered through the Definition phase (Chapter 3), either by reference to a pre-existing project charter or by explicit statement of objectives.

Another purpose served by the Definition phase is to ensure shared understanding of what is meant by the term "risk." This is important since, as we have already seen (Chapter 2), there are several definitions in current use in the risk management community, and the project team need to know which one they are using when seeking to identify risks to their particular project. If, as recommended earlier, a broader definition of risk is used, which explicitly includes both opportunity and threat, then this will be stated in the Definition phase, and the remainder of the risk management process can be implemented in the light of this understanding.

Having done the essential groundwork in the Definition phase to initiate the risk process on a firm foundation, the project can now move on to the first step in the cyclic part of the risk process, namely Risk Identification. This chapter gives guidelines on how to ensure that good intentions to identify opportunities alongside threats are translated into reality, and presents a number of techniques that can be used within the risk process to capture both opportunities and threats together.

IDENTIFICATION CHALLENGE 1: IDENTIFYING ALL RISKS?

The objective of Risk Identification is to expose and capture details of as many risks as possible, and to do this proactively, in advance of them occurring, to give the project team enough management space to deal with the risks before they might happen. It should of course be accepted that it is never possible to identify all risks that a particular project might face, for a number of reasons:

Some risks are inherently unknowable or unforeseeable, being the product of random chance.

Some are "future risks," uncertain events or combinations of circumstances that have not yet occurred and that cannot be predicted from the perspective of this point in time.

Some "emergent risks" will arise only as a result of actions to be taken in the future, and since those actions have not yet occurred the risk may be hidden.

Some risks arise from the decisions or choices of "invisible stakeholders," i.e., those people or parties with the ability to affect the project but who are hidden from or unrecognized by those trying to identify risks.

Some risks are "perceptually concealed" from those seeking to identify risks; i.e., the people are unable to perceive the risk due to their own psychological or emotional biases or paradigms—the risk is literally inconceivable to them.

There is a special case of the "perceptually concealed" type of risks, which can present a major hindrance to Risk Identification. This derives from the understanding of the definition of a risk itself, and is of particular relevance when the aim is to include management of opportunities within the risk process. Where participants in the Risk Identification process perceive risk as equating to threat and do not see opportunities as included within the definition of risk, then it is inevitable that Risk Identification will fail to identify those opportunities that may exist. This should be addressed by the Definition phase (Chapter 3), where the definition of risk to be used is clarified, but there remains the danger that opportunity identification is a significant blind spot in the Risk Identification process.

Nevertheless, the task of the Risk Identification phase is to identify as many risks as possible with sufficient lead time to enable them to be managed effectively. The fact that some risks cannot be identified from the current perspective in time (i.e., "future risks" and "emergent risks") should encourage the Risk Identification task to be repeated at regular intervals throughout the project (to spot those risks that were previously hidden but that have now become visible). The limitations of trying to identify risks from one person's limited perspective can be overcome by involvement of a wide group of project stakeholders in Risk Identification so that as many perspectives as possible are covered (overcoming the problems of "invisible stakeholders" and "perceptually concealed risks"). And the conceptual blind spot that prevents opportunities from being identified needs to be tackled by both definition and process, both ensuring that participants in Risk Identification know that opportunity identification is expected, and providing techniques and tools that facilitate identification of opportunities alongside threats.

IDENTIFICATION CHALLENGE 2: WHEN IS A RISK NOT A RISK?

There is another challenge faced by those seeking to identify risks, which is to ensure that the Risk Identification phase identifies risks. While this may sound axiomatic, in fact it is common for Risk Identification techniques to result in other things being exposed and recorded, which are not risks. For example, the project team may identify *issues* currently faced by the project that are causing problems, or they may wish to record *difficulties* they are facing in executing agreed tasks, or they may try to discuss *political, organizational, or personnel matters*. Often there is no other mechanism for these things to be raised, so project team members use the risk process to make management aware of things that do not fall under the definition of risk—i.e., they are either not uncertain events, or they do not affect project objectives.

Clearly it is important for the risk management process to remain focused on dealing with risks if it is to be fully effective. If the output of Risk Identification techniques includes items that are not risks, it dilutes the value of the risk process, introduces distractions at the early stages, obscures genuine risks, and can lead to frustration and disillusionment among participants. We cannot afford to allow the process to be diluted or distracted at the outset by inclusion of nonrisks.

Even worse is the situation where nonrisks are identified during the risk process, but they are not distinguished from risks. In other words, the situation often arises where the project team identify things that are not risks, but they think they are risks, and record them as such. This is a problem because the risk process is designed to deal with risks, i.e., with uncertainties that can affect objectives. When nonrisks are mistakenly seen as risks, the risk process can become discredited since it is apparently "failing" to deal with these "risks." This can lead to loss of momentum or commitment to proactive management of risk by the project team, resulting in unmanaged risks and further failure. This vicious circle must clearly be avoided.

In fact the most common error made during Risk Identification is to fail to distinguish between *causes* of risk, genuine *risks*, and the *effects* of risks. These are defined as follows:

Causes: Definite events or sets of circumstances that exist in the project or its environment, and that give rise to uncertainty. Examples include the requirement to implement the project in a developing country, the need to use an unproven new technology, the lack of skilled personnel, or the fact that the organization has never done a similar project before.

Risks: Uncertainties that, if they occur, would affect the project objectives. Examples include the possibility that planned productivity targets might not be met, interest or exchange rate fluctuations, the chance that client expectations may be misunderstood, or whether a contractor might deliver earlier than planned.

Effects: Unplanned variations from project objectives, either positive or negative, that would arise as a result of risks occurring. Examples include being early for a milestone, exceeding the authorized budget, or failing to meet contractually agreed performance targets.

The relationship between cause, risk, and effect is illustrated in Figure 1. This shows the most simple case of one cause leading to a single risk that in turn could have just one effect. Of course, reality is considerably more complex than a set of one-to-one links, and each relationship (cause-risk and risk-effect) can be one-to-many, many-to-many, or many-to-one, creating a complex web of interrelationships. The position is further com-

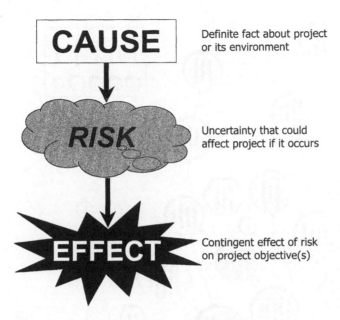

Figure 1 Distinguishing cause, risk, and effect.

plicated by the recognition that risks if they occur can become causes of new risks, and effects can also become causes starting additional chains, thus propagating the cause/risk/effect network into further complexity. A more detailed illustration of this is given in Figure 2, which shows a Risk Concept Map (an alternative format for the cause/risk/effect diagram, in which the three elements are called risk driver/risk situation/impact). While this more complex representation is doubtless closer to reality, the simplification of Figure 1 is sufficient to illustrate the ideas for the purposes of this discussion.

It is important to maintain a clear separation between cause, risk, and effect during Risk Identification. Causes of risk are definite features of the project, and as such they are not uncertain. If they are mistakenly labelled as "risks," they will distort the risk assessment phase of the process, since their probability of occurrence is certain (they are facts). This means that those performing risk assessment will rank causes higher than genuine risks, since causes have a "probability of occurrence" of 100%, while true risks have a lower probability i.e., < 100%. This obscures genuine risks, which may not receive the appropriate degree of attention

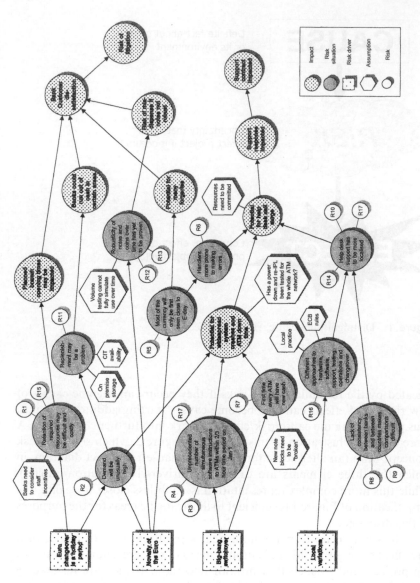

Figure 2 Risk concept map. (From Bartlett, 2002.)

they deserve. The credibility of the risk process is also undermined, since it is often not possible to tackle causes effectively and it appears that so-called "risks" are not being managed.

Similarly, if effects are included as "risks," they will distract attention from genuine risks. Effects are contingent events that will not occur unless risks happen, so they are unplanned potential future variations. The aim of risk management is to tackle genuine risks, and if risk management is working properly, effects will not be experienced. As effects do not yet exist, and indeed they may never exist, they cannot be managed through the risk management process. As with causes, inclusion of effects as so-called "risks" will divert attention from genuine risks and will damage the credibility and effectiveness of the risk process.

A risk is defined as any uncertainty that, if it occurs, will affect project objectives. Risks therefore have two dimensions, namely the uncertainty (described as "probability") and the effect on objectives (described as "impact"). Only genuine risks have both of these dimensions, since causes are certain and have no associated "probability" dimension, and effects are simply a description of the "impact" dimension of a risk.

Using Metalanguage to Describe Risks

To assist participants in the Risk Identification phase to clearly separate risks from their causes and effects, an approach is required that forces a distinction. Use of an appropriate *metalanguage* can provide structure to the description of risk in a way that achieves this separation. Metalanguage uses words or symbols to structure thinking and describe a concept in a systematic and formal way.

Risk metalanguage provides a three-part structured "risk statement" that includes cause, risk, and effect. By requiring each element to be explicitly stated using precise words, confusion between the three is minimized.

The three elements of the risk metalanguage can be summarized in the following statement:

"As a result of < *definite cause* >, < *uncertain event* > may occur, which would lead to < *effect on objective(s)* >."

Examples include the following:

1. "*As a result of using novel hardware, unexpected system integration errors may occur, which would lead to overspending on the*

project." The use of novel hardware is a definite fact that is a cause of uncertainty and gives rise to risk. The risk itself is the possibility of encountering unexpected errors, which of course might not occur. If, however, that were to happen, there would be an effect on the project budget. A proactive approach can be adopted if the risk is correctly identified as "There may be an unexpected level of integration errors." Planned actions might then include more rigorous preliminary checks or prototyping, use of experienced personnel, or creation of more float for this activity. If, however, "The required use of novel hardware" was wrongly identified as a risk, it might be hard to develop suitable responses, since this fact forms part of the project requirement. Similarly, if "Project overspend" was thought to be the risk, no response would be possible since this would only occur if something else happened.

2. "*Because our organization has never done a project like this before, we might misunderstand the customer's requirement, and our solution would not meet the performance criteria.*" The definite factual cause is the organization's lack of relevant prior experience. The uncertain event is the possibility of misunderstanding requirements, although lack of experience does not necessarily mean that this will happen, so it is a risk. The contingent future effect would be failure to meet the performance objective.

3. "*We have to outsource production; we may be able to learn new practices from our selected partner, leading to increased productivity and profitability.*" Here the cause of the uncertainty is the requirement to find a production partner, with the upside risk (opportunity) that process improvements might be shared, leading to a positive effect on productivity and profitability.

4. "*The plan states a team size of ten but we only have six staff available, so we might not be able to complete the work in the required time, and we could be late.*" The definite staff shortage is the cause, giving rise to a risk that the available team may be too small for the required scope, with a possible effect on project timescales.

It is possible to identify keywords for each of the three elements of the risk metalanguage, whose use indicates correct separation of cause from risk from effect, though the inherent flexibility of the English language (and the lack of grammatical rigor on the part of most speakers) means that any word-based rules cannot be infallibly applied. For example:

> *Causes* are definite facts or conditions about the project, so their element of the risk metalanguage includes present tense phrases such as "*is, do, has, has not, are...*"
>
> *Risks* represent uncertain future events that may or may not occur, and are described with words like "*may, might, could, possibly...*"
>
> *Effects* are conditional future events, contingent on the occurrence of risks. These are the hardest to describe unambiguously because of the imprecise nature of the English language. The subjunctive form "*would*" is the proper way to describe effects, but a simple or conditional future tense is often used, namely "*will, could...*"

The use of this type of risk metalanguage should help to ensure that the Risk Identification process actually identifies risks, rather than anything else. Without this discipline, Risk Identification can produce a mixed list containing risks and nonrisks, leading to confusion and distraction later in the risk process.

A modification of the risk metalanguage described above can be used to specifically expose uncertainties, based on consideration of strengths or constraints. Organizational strengths can be identified using SWOT Analysis (described later in this chapter), and a metalanguage construct can use these as potential causes of opportunities, thus:

> "Because we have < *strength* >, we might be able to create/exploit < *opportunity* >, which would lead to < *benefit* >."

Similarly, Constraints Analysis (also described below) can lead to testing of constraints via the risk metalanguage:

> "If we could relax or remove < *constraint* >, we might be able to create/exploit < *opportunity* >, which would lead to < *benefit* >."

SEEING WITH TWO EYES

The challenges outlined above affect Risk Identification in all types of risk process, whether the process is threat-focused or whether the goal is to deal with opportunities explicitly within a common process alongside threats. Clear thinking is important during this Risk Identification phase if there is a serious intention to include opportunities in the risk management process as well as threats. And this must find expression in the tools and techniques used for Risk Identification, which must be capable of exposing both opportunities and threats equally effectively.

Of course, many techniques are already commonly used to identify risks, although in practice most of the time these are used exclusively to expose threats to project objectives. The 16th-century Dutch Renaissance humanist scholar Desiderius Erasmus (?1466–1536) commented that "It is well known that among the blind, the one-eyed man is king." Applying this to Risk Identification, the message is clear. The uncertainty surrounding all projects includes both upside opportunities and downside threats. As commonly practiced though, Risk Identification usually only addresses threats, i.e., those risks with the potential to affect project objectives negatively. While this may be seen as a major weakness and shortcoming in the traditional approach to risk management, it is surely better than nothing. The ability to look ahead and see potential dangers that could have a harmful or adverse impact on one or more project objectives is a huge advance over the reactive approach adopted by some project managers, who seem content to "wait and see what happens," then deal with the inevitable crisis later.

It is certainly better to identify only threats than not to identify any risks at all. However, in the terms of Erasmus, this represents the situation of the one-eyed man when compared with the blind. Best of all is to be able to see with both eyes, and to use them both equally. To continue the metaphor a little further, monocularism leads to distortion, blind spots, and lack of perspective; the ability to combine the view from two eyes creates true perspective, depth of field, and the ability to make accurate assessments of range and proximity.

There are two ways to progress from being "one-eyed" to seeing with two eyes.

- The first is to open the other eye and use both together, i.e., becoming a two-eyed man. This means trying to adapt those "one-eyed" techniques commonly used for Risk Identification but that generally tend to concentrate on threats, and seeking to broaden their application to include opportunities.
- Second, one might seek the assistance of someone who is "one-eyed" in the other direction, in other words looking for different techniques that are intrinsically directed toward opportunity

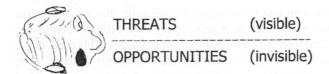

THREATS   (visible)

OPPORTUNITIES   (invisible)

The one-eyed man

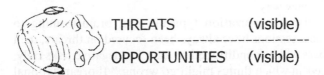

THREATS   (visible)

OPPORTUNITIES   (visible)

The two-eyed man

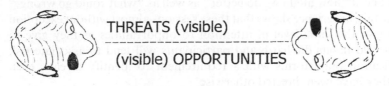

THREATS (visible)

(visible) OPPORTUNITIES

Two one-eyed men

Figure 3 Different perspectives on risk.

identification. The combined vision of two one-eyed men with different perspectives may ensure a more complete view, allowing both opportunities and threats to be identified.

These two approaches are illustrated in Figure 3, and are discussed in the following sections.

ADAPTING COMMON IDENTIFICATION TECHNIQUES

Risk identification uses several techniques that in themselves are not limited to exposing threats, but could equally well be used for revealing opportunities (in other words, seeking to "open the second eye," using the accepted familiar approach to identify both types of risk). There is, however, a problem with trying to adapt a technique that project teams have been using for some time in a particular way. Human beings are creatures of habit, and tend to follow patterns of behavior. This tendency is strong and hard to resist, even when a conscious decision is made to do otherwise. So when project team members are encouraged to use a familiar Risk Identification technique in an unfamiliar way, there is a natural tendency to revert to how things were done previously, even if mental assent is given to the new way.

For example, if an organization typically uses brainstorming to identify risks, but these sessions have always been focused on threats, then project team members will be conditioned to look for negative risks only, thinking about "ways in which things might go wrong." If organizational policy changes and risk is defined to include opportunity as well as threat, then those attending risk identification brainstorms might be encouraged to talk about "how might we do better" as well as "what could go wrong." Despite this, experience shows that these sessions almost without exception fail to give the same level of attention to opportunities as they give to threats. Old habits die hard, and participants will tend to associate risk identification brainstorms with a requirement to identify threats, even when they have been directed otherwise.

Nevertheless it is possible to use the common Risk Identification techniques to identify both opportunities and threats together, as long as there is awareness of the tendency to revert to the way things have been done previously. This means that proactive management is necessary to ensure that the techniques are applied with the required breadth of scope. The most frequently used Risk Identification techniques are outlined briefly below, discussing how they might be modified to incorporate opportunity identification alongside threats.

Brainstorms

Brainstorming is the most commonly used method of Risk Identification. It is popular because it is a group creativity technique, giving a number of side benefits in addition to its primary purpose of identifying risks. For example, it can be a very powerful means of team building, giving the project team an opportunity to work together and discuss matters that are important to the success of their project. It creates shared ownership over the issues identified, since the whole group present at the session buys into the outputs and has an emotional commitment to support the results of the brainstorm. As an exercise in creative thinking, brainstorming can also be great fun, since one of the basic rules is that anything can be said without criticism or judgment. This also provides a platform for people to express their concerns openly without fear of being blamed or criticized (although an adversarial organizational culture can preclude this openness, leading to what is sometimes known as "blame-storming"!). The brainstorm session is also seen by senior management as a valuable chance to contribute to the project and meet the team, although it can sometimes be misused by senior managers to display their wisdom and experience, or to exert undue influence over the project.

These benefits are all related to the human aspects of brainstorming, and they reveal two key factors required for a successful brainstorm session:

> *Attendance*—it is important to get the right people to attend the brainstorm session. Particularly when using the brainstorm for Risk Identification, attention must be paid to inviting key project stakeholders, representing all parties with an interest in the project. It is often the case that those individuals who know most about the project and its risks are never invited to brainstorm sessions, perhaps because they are not prominent in the hierarchy and all the places in the brainstorm have been taken by senior managers keen to make their contribution publicly.
>
> *Assistance*—given the potential for significant group dynamics effects during a brainstorm session, the role of the facilitator is crucial to ensuring a successful outcome. This person should be familiar with the project and with the risk process, and should be able to manage the interpersonal dimensions within the group to counter any tendency toward "blame-storming" or criticism. The facilitator needs to be able to encourage quiet people to make their contribution, while channeling and controlling the more talkative participants. They must also be sensitive to issues of power, status, and position, and react appropriately to ensure that all voices are heard.

Brainstorming was originally developed as a problem-solving technique, seeking to create a framework where the team could propose possible solutions and generate ideas for further investigation and development. The classical approach to brainstorming has four rules:

1. Creativity and freethinking is welcomed and encouraged.
2. Quantity is required, since the more ideas are generated, the greater the chance is of finding a solution.
3. Criticism is ruled out, with judgment being deferred to outside the session.
4. Combination and improvement are sought, to encourage the generation of better ideas by building upon the ideas of others.

Brainstorming for Risk Identification follows the same basic approach, but instead of problem solving or idea generation, the task of the brainstorm session is to identify possible risks to the project objectives. This requires some modifications to the classical brainstorm, although the basic four rules still apply. The main change is to introduce structure to a risk identification brainstorm, rather than allowing a completely free session. The danger of unstructured brainstorming is that the team may fail to consider risks in some areas of the project, spending all the brainstorm time concentrating on a subset of the project scope. The solution is to hold a semistructured brainstorm, for example using the project Work Breakdown Structure (WBS). Here the team are asked to brainstorm risks within a WBS major area for a limited period of time (say 10–20 min), before moving on to the next WBS area. This approach allows all areas of the project to be covered during the brainstorm. An alternative is to structure the brainstorm using risk categories or a Risk Breakdown Structure (RBS) as described in Chapter 5, ensuring consideration of all potential sources of risk to the project.

Given the right attendance, a good facilitator, some structuring, and adherence to the accepted rules of brainstorming, one may ask whether anything else is required to enable a brainstorming approach to be used for opportunity identification as well as exposing threats to the project. Surely it should be enough for the facilitator to simply announce at the outset that this brainstorm session will be used to identify risks with upside impact as well as those with downside?

While in theory this should be enough, the influence of habit mentioned above tends to mean that brainstorm participants will still produce mainly threats, despite being invited to identify opportunities as well. The effect of previous experience is particularly strong in a creative environment, when a person is responding intuitively without relying heavily on rational thought (this effect is governed by the operation of heuristics, which are discussed in detail in Chapter 9). Human nature is also a factor, as it seems that people involved in projects prefer to criticize rather than to praise, and it is easier to find fault than to be constructive. Similarly in risk identification brainstorms, the natural tendency seems to be to seek the negative to the exclusion of possible opportunities. (This may of course be cultural, particularly in Anglo-Saxon countries and organizations where skepticism seems to be a way of life! The influence of national culture on risk perception is also discussed in Chapter 9.)

The tendency of brainstorms to concentrate on threats can be easily countered by introducing an explicit focus on opportunities into the session structure. If the brainstorm is structured around the WBS or risk categories as suggested above, the facilitator can simply divide consideration of each element into two parts, one dealing with opportunity and another with threat. The use of de Bono's "Six Thinking Hats" technique to structure brainstorm workshops might also facilitate this type of approach, focusing the group in turn on creative thinking (green hat = new ideas), logical positive thinking (yellow hat = opportunity), logical negative thinking (black hat = threat), etc. When brainstorm sessions are structured in this way to consider both upside and downside, the advantages of brainstorming are retained, but the structure actively encourages opportunity identification alongside the threats that participants tend to identify more readily.

Another alternative is to divide people into two separate brainstorm sessions, with one session tasked with identifying threats and the other looking for opportunities. The two groups can then meet together to review each other's outputs and to create a combined list of threats and opportunities. This approach can be refined further by forming the two brainstorm groups taking into account the personality types of participants, where this information is known and understood. The idea here is to use optimists to spot opportunities and pessimists to find threats, playing to the psychological preferences of each personality type. While this may result in heightened focus on each type of uncertainty, there is the danger of overemphasis on one side or the other, with the optimists becoming unrealistically positive about prospects for exceeding project objectives, while the pessimists get extremely gloomy and negative. Holding a plenary session should counter this tendency, allowing optimists to review the threats, and pessimists to check the opportunities, hopefully leading to a realistic identification of both types of uncertainty.

Checklists

A risk identification checklist is a powerful means to ensure learning from previous experience, since it captures risks that have been identified on previous projects, as well as risks that were not previously identified but that occurred. The approach is usually to present each item as a simple closed question, allowing only one of a few specified answers, namely Yes/No/Unknown/Not Applicable. Questions are usually phrased positively, such that an answer of "No" or "Unknown" indicates a possible risk to the project. For example, the question might be "Is the requirement well understood?" or "Are all subsystem interfaces defined and acceptable?" An answer of "No" to such a question reveals a definite source of risk, whereas "Unknown" indicates an area of uncertainty that may pose a risk and should therefore be explored in more detail.

The list of questions in the risk identification checklist can be structured using the WBS, or in some other way, for example by sources of risk or using an RBS (see Table 1 in Chapter 1 for example).

It is also possible for the checklist to include potential responses to risks that have been found to be effective on previous projects, thus allowing the checklist to be used for two purposes: Risk Identification and Response Planning.

Risk identification checklists are popular for several reasons. The chief of these is that they are extremely easy and quick to use. They can also be used by an individual without needing to call together the project team. Of course, both of these advantages are also potential shortfalls. The ease of use can lull the person into a false sense of security, believing that simply running through the checklist will identify all possible risks on the project—resulting in a literal "tick-in-the-box" mentality. And the ability to work without input from other team members can result in a limited perspective that is subject to personal bias or preconception, and that can therefore fail to identify key risks.

Another advantage of using checklists for Risk Identification is the link with previous projects, facilitating learning from experience. In some industries, the professional bodies and associations maintain checklists of generic risks affecting projects in their sector, and these can be obtained freely or for a nominal cost. Examples exist in the defence, construction, IT, engineering, software development, offshore oil and gas, and pharmaceutical sectors. One project has even recently been concluded that attempted to identify a "Universal Risk List" (a joint collaboration during 2002 between the Risk Management Specific Interest Group of the Project

Management Institute, PMI Risk SIG, and the Risk Management Working Group of the International Council On Systems Engineering, INCOSE RMWG). This project originally set out to produce a checklist of "universal risks" that might affect any project in any sector anywhere in the world, although this ambitious goal was eventually downgraded to listing generic "risk areas" (see Table 1-1).

While generic checklists from professional associations provide access to the experience of many projects in a specific industry, there is perhaps greater value to an organization in producing its own risk identification checklist. This can capture specific risks to which projects within the organization have been exposed, allowing a more targeted Risk Identification. This also allows the possibility of greater openness, since an internal risk checklist can be scrupulously honest, compared to those produced for public consumption, which may have been sanitized to remove items of commercial sensitivity or political embarrassment.

A number of dangers in using checklists must be addressed:

1. Checklists capture previous experience on past projects, listing those risks that have been identified or have occurred before. While the value of this is self-evident, there is an implicit assumption that the checklist represents a complete statement of all possible risks that might affect the current project. This is almost certain to be a false assumption, since every project is unique in some way. As a result, there are also likely to be some unique risks facing each project, which by definition will not appear on any checklist compiled from past experience.

2. Generic risk identification checklists claim to apply to any project in the given sector, and they bring together risk data from several or many projects. Users of such checklists must, however, question whether the data set on which the checklist is based is relevant to the project under consideration. If the questions in the checklist are not drawn from similar projects, their value and applicability will be limited. In the extreme case, such checklists may even be misleading.

3. Checklists are produced by adding risks from the postproject review process of completed projects. This of course assumes that such a process exists and is used effectively. Project closure and postproject review are perhaps the weakest parts of the entire project management process in most organizations. As a result there is no guarantee that risks are being captured from previous

projects and being included in checklists. The completeness and
value of the checklist will be in direct proportion to the effectiveness of the postproject review process.

4. Given that a process exists to learn lessons about risk from
    completed projects, and that risks will be added to a checklist
    (either generically by a professional body, or specifically within
    an organisation), there is a requirement for checklists to be main-
    tained. For example, a middle-sized project-based organization
    may complete 15–20 significant projects each year, and each
    project might be expected to identify 5–10 new risks that were
    not previously identified on past projects. If all of these new
    risks were to be added to the organization's risk identification
    checklist, the list would grow by 75–200 risks each year! Clearly
    the checklist would become unmanageably large in a short pe-
    riod if it were allowed to grow uncontrolled at this rate. As a
    result is it essential that the checklist should be maintained,
    pruning out risks that are no longer relevant, as well as adding
    new ones.

5. The ability to relate risk data from previous projects to the
    current one requires use of a common language or classification
    system enabling risks to be mapped easily between projects.
    Often projects within an organization have no accepted way of
    describing or grouping risks. This can hinder the lessons learned
    process, as well as making it difficult to create and use a checklist
    for Risk Identification.

Despite these shortcomings, checklists remain a popular tool for
identification of risks. Given this popularity and the undoubted strengths
of checklists to facilitate learning from previous experience, can this
technique be used to identify opportunities as well as threats?

Again as for brainstorming, the answer is yes in principle, but no in
practice. There is no inherent reason why a checklist should not capture
opportunities found on previous projects, and pose questions to future
projects about whether these same opportunities might apply. However,
given the relatively recent inclusion of opportunity management within
the risk process, most historical risk identification checklists contain only
threats. Their use for future projects therefore perpetuates the situation
where the only type of risk identified is the threat to project objectives, with
opportunities being overlooked. This will continue until opportunity
management becomes better established as an integral part of the risk

process, at which time opportunities will naturally be incorporated into risk identification checklists alongside threats.

For the organization or project manager for whom the checklist is a preferred means of Risk Identification, the solution is to begin to modify checklists to include opportunities, reviewing the experience of projects where opportunity management was practiced, and building that learning into the checklist to make it available to future projects. The checklist will then begin to include opportunities that were identified before, whether they were realized or not, and will prompt project managers to ask whether these opportunities might also apply to their current project. This exercise should also draw on the lessons-learned process to identify positive elements of previous projects, and include items in the checklist that ask "This went well before—could we do the same again? (or even better?)."

In this way, a checklist can be created that explicitly includes possible opportunities as well as potential threats, and can then be used to support a broad identification of the full range of risks facing a project. This will, however, take time to generate, and depends on the organization's commitment to proactive opportunity management as well as the effectiveness of the lessons-learned process.

Interviews

One of the most effective means of identifying risks is to talk to people. As a result, the risk interview has become established as an accepted method of Risk Identification. As for other Risk Identification techniques discussed above, structure is a valuable aid to the risk interview. The technique can of course be implemented very simply, just sitting down with key project stakeholders and asking them what they perceive as risks to the project. Or it can be done in a more structured way, using an interview structure based on the WBS or risk categories or RBS. The structured approach is preferable since this is more likely to expose risks across the broad scope of the project than a casual chat about areas of concern.

A good project manager should be aware of the majority of risks faced by the project, based on familiarity with the project charter and/or business case, its objectives, constraints, and assumptions, and the organizational context within which the project operates. The experience of some risk practitioners has led to an estimate that the project manager might be able to identify 60–80% of foreseeable risks through use of a structured interview approach. The areas where the project manager might not be able to identify risks are those requiring specialist knowledge, for example technical risks or commercial risks. To fill the gap in these areas, interviews might also be held with one or more technical authorities on the project, familiar with the details of the technology involved on the project, as well as interviewing a representative of the contracts or commercial department, able to identify specific risks relating to this aspect of the project. One might expect specific technical risks outside the knowledge or experience of the project manager to total 10–30% of the total number of foreseeable risks, and there may be an additional 5–10% of risks that specifically relate to commercial or contractual issues. These figures suggest that interviewing the project manager plus specialists in the technical and commercial aspects of the project can expose the majority of foreseeable risks on a project, as illustrated in Figure 4.

Of course other factors are required for a successful risk identification interview. It is not possible to arrange an interview time and turn up expecting risks to be identified effortlessly. Preparation is essential, on the part of both the interviewer and interviewee. Both need to have reviewed

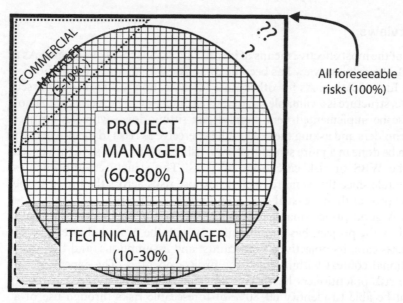

Figure 4   Scope of risks exposed during risk interviews.

and be familiar with the project objectives and status. The interviewee should have spent some time prior to the interview considering possible risks, to make best use of the time available for the interview. Similarly the interviewer should do some homework, preparing a prompt list to be used during the interview to assist the interviewee, and thinking about leading questions to facilitate discussion.

Another essential ingredient for a successful risk identification interview is trust. The interview should be conducted in a confidential environment, where the interviewee can be encouraged to express concerns honestly without fear of reprisal or blame. This requires an open relationship between interviewer and interviewee, with mutual respect and a shared commitment to work together to expose the risks facing the project. Existence of such a relationship is not automatic, and the risk interviewer may need to work on developing trust, through use of open questions, lack of a judgmental or critical attitude, demonstration of respect, and recognition of confidentiality.

There are also specific interview skills that are required for a successful risk identification interview, including active listening and selective questioning. Active listening requires the interviewer to play a part in the process even when the interviewee is speaking. The characteristics of active listening can be described using the ACTIVE mnemonic:

> Attention—the listener must demonstrate that attention is being paid to the speaker, through body language, eye contact, reflection of what is said, etc.
>
> Concern—it must be clear that the subject being talked about matters to the listener, rather than there being a detached impersonal approach.
>
> Time—listening cannot be rushed, and the listener must allow the speaker to take whatever time is required to express himself or herself.
>
> Involvement—the listener should be involved with the speaker, giving the impression of a two-way conversation even if there is only one speaker.
>
> Vocalization—short expressions indicating that the listener is hearing what is said will encourage the speaker, such as "I see ... that makes sense ... of course ..."
>
> Empathy—the speaker will be encouraged to open up and continue expressing his opinion when the listener shows that the speaker's feelings on the subject are shared.

Selective questioning requires appropriate use of different question types to ensure that the interview maintains momentum and moves toward a conclusion. Question types traditionally include:

Open questions, to gather information and facts
Probing questions, to gain additional detail
Hypothetical questions, to suggest an approach or conclusion
Reflective questions, to check understanding
Closed questions, to bring agreement, commitment, and conclusion

An effective interviewer will use these different types of question during the interview to provide structure, direct flow, and reach closure as appropriate, using the "question funnel" approach shown in Figure 5. The order of

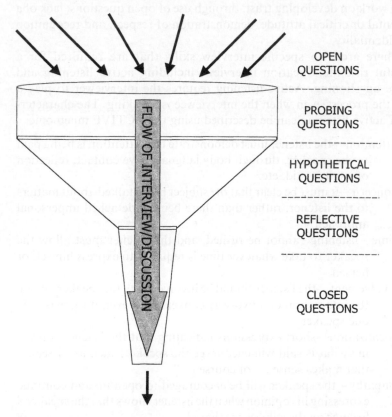

Figure 5   The question funnel.

questions usually starts with open and ends with closed, though not all types are required in every situation. Some interviewers adopt a creative approach to questioning, using other nontraditional question types, such as "fantasy questions" (e.g., "If you dreamed about your project and it turned into a nightmare, what would be happening?" or "I am your fairy godmother and have granted you three wishes to use on your project—what will you do first?"), or deliberately naïve questions (e.g., "What advice would you give a fresh graduate trainee joining the project to help him keep out of trouble?" or "I really don't understand why this project shouldn't be a huge success—please explain").

As with the other traditional techniques for Risk Identification, interviewing is a powerful tool when used properly, and can result in exposure of the major risks to a project. There are significant benefits when using interviews to identify risks, including:

> The positive effect on morale when people feel they are being listened to and that their opinion matters
>
> The ability to explore risks in some detail, spending time during the interview to test understanding and make sure that the interviewer has captured the essence of the concern
>
> Encouragement to identify risks in a confidential one-to-one environment that might otherwise remain hidden when using group-based techniques that are more public or where the originator of identified risks is visible

Of course, risk interviews can be used to explore and identify opportunities as well as threats. This, however, depends on the interviewer and interviewee sharing a common understanding that risk includes both opportunity and threat, and also requires them to remember this during the detail of the interview. Like other techniques, risk identification interviews suffer from the tendency to revert to the familiar focus on threats if there is not an active and sustained effort on the part of both interviewer and interviewee to include opportunities consciously within the scope of the interview.

NOVEL TECHNIQUES FOR IDENTIFYING OPPORTUNITIES

Although it is possible to modify use of the existing familiar techniques for Risk Identification to identify opportunities, there are problems as outlined above, where force of habit often inhibits use of these approaches to identifying anything other than threats, and identification of opportunities is therefore blocked.

As a result, organizations may find it more effective to introduce new Risk Identification techniques as part of the risk process alongside the more familiar ones, with the new techniques being focused specifically toward identifying opportunities. This section presents several such techniques that are particularly suitable for opportunity identification, and that are mostly adaptations of existing approaches from other disciplines outside risk management. The techniques outlined here are SWOT Analysis, Constraints Analysis, and Force Field Analysis, though there are doubtless others that could be adopted and adapted in a similar way. It is recommended that these should be used together with the traditional Risk Identification techniques, perhaps expecting the familiar approaches to produce mostly threats, while the new techniques might generate mostly opportunities. In combination a broad identification of both upside and downside risks is likely to be achieved.

SWOT (Strengths/Weaknesses/Opportunities/Threats) Analysis

It is perhaps unsurprising that SWOT Analysis should be the first technique offered here as a suitable approach for explicit identification of opportunities during the risk process. After all, the abbreviation "SWOT" reveals that at least a quarter of the focus of this technique concerns Opportunities, along with Strengths, Weaknesses, and Threats.

These four elements of a SWOT Analysis can be defined as follows :

1. A *Strength* is a characteristic, resource, or capacity the organization can use effectively to achieve its objectives.
2. A *Weakness* is a limitation, fault, or defect in the organization that might keep it from achieving its objectives.
3. An *Opportunity* is any potentially favorable situation in the organization's environment that may positively affect the project that is being undertaken. It may include a trend or change of some kind, or an overlooked need that increases demand for a product or service.
4. A *Threat* is any potentially unfavorable situation in the organization's environment that may be damaging to the project that is being undertaken. The threat may be a barrier, a constraint, or anything external that might cause problems, damage, or injury.

SWOT Analysis is usually conducted in a workshop setting with key members of the project team and/or other stakeholders. As such, it could be regarded simply as another way to structure a risk identification brainstorm, since the SWOT Analysis session is usually run as a facilitated creative group exercise implementing brainstorming rules, but structuring the session into four parts, one each for Strengths, Weaknesses, Opportunities, and Threats.

This seems simple enough, but there are at least two problems with treating SWOT Analysis merely as a four-part brainstorm.

First, there is often confusion over the difference between Strengths/ Opportunities and Weaknesses/Threats, with participants in the workshop not being able to clearly distinguish between them.

Second, so-called "SWOT Analysis" sessions are usually only doing "SWOT Identification," with little or no analysis of the results.

Taking these in turn, a clear distinction should be made between the various SWOT components. Strengths and Weaknesses relate to the organization itself, describing characteristics and resources that the organization brings to the task at hand, answering the question "Who are we?" Opportunities and Threats, on the other hand, relate to the specific project being considered, and are about "What are we doing?"

Second, it is important to consider the effect of Strengths and Weaknesses on Opportunities and Threats. It is obvious that "who we are" affects "what we do," since the level of risk faced by different organizations performing the same project is different. The analysis part of SWOT Analysis requires the relationship between Strengths/Weaknesses and Opportunities/Threats to be determined and assessed. If this level of analysis is not performed, then the so-called SWOT Analysis has only achieved a four-part brainstorm, without analyzing the implications of the interactions between the various SWOT elements.

The following guidelines indicate how a SWOT Analysis session might be conducted as an effective Risk Identification technique, and include a mechanism for assessing the effects of Strengths/Weaknesses on Opportunities/Threats. (It should be noted that the approach recommended here includes elements of risk assessment, which can either be included within the SWOT Analysis session or conducted separately.)

SWOT Analysis is best undertaken in a facilitated workshop setting, attended by project team members with appropriate knowledge and expertise. The purpose is to assess the riskiness of a project, cross-analyzing opportunities and threats against current strengths and weaknesses in the organisation. The workshop is usually run with two distinct elements:

1. Identify and rate opportunities and threats
2. Determine influence of strengths and weaknesses

Results can be documented on a SWOT Worksheet (see Fig. 6 for an example format), which can be implemented as an electronic spreadsheet to calculate scores automatically.

The first element of the SWOT Analysis workshop is to identify opportunities and threats that might affect achievement of the project objectives. Opportunities are uncertainties that might have a beneficial effect if they occur, whereas threats are unplanned uncertain events that could have an adverse effect. Opportunities and threats can be identified by any technique, although structured brainstorming is most common. A consolidated list of opportunities and threats is produced, taking care to express these as uncertainties (note that in some cases the same issue may be expressed both positively as an opportunity and negatively as a threat— both should be included separately).

Each item in the consolidated list of opportunities and threats is then assessed by workshop participants. For example, the probability (P) of the opportunity or threat occurring can be scored as follows:

    0 = not possible
    1 = unlikely
    2 = possible
    3 = probable

The impact (I) of each opportunity or threat can also be scored as follows, with opportunities having positive scores and threats having negative scores:

    0 = no impact
   ±1 = minor impact
   ±2 = significant impact
   ±3 = major impact

A P-I Score can then be calculated for each opportunity or threat by multiplying scores for probability (P) and impact (I), producing a P-I Score in the range 0 to +9 for opportunities, and 0 to −9 for threats. Scores are recorded on the SWOT Worksheet.

PROJECT : _____

SWOT WORKSHEET
(gray cells calculate automatically)

WORKSHOP DATE : _____

		O/T rating			Strengths			Weaknesses			Modified O/T Score	Total O/T Score
		Prob (P)	Impact (I)	P x I Score	S1	S2	S3	W1	W2	W3		
Opportunities	Opportunity 1	1	2	2	2		2			-2	2	21
	Opportunity 2	2	1	2		2		-2			2	
	Opportunity 3	3	3	9							9	
	Opportunity 4	2	3	6	2						8	
	Opportunity 5	1	0	0							0	
Threats	Threat 1	3	-2	-6	2		2	-2	-2		-6	-15
	Threat 2	3	-3	-9							-7	
	Threat 3	1	0	0							0	
	Threat 4	0	-1	0							0	
	Threat 5	1	-2	-2		2		-2			-2	

Figure 6 SWOT Worksheet.

The second element of the workshop is to identify existing strengths and weaknesses for the organization in relation to the project. These detail characteristics of the organization as a whole, and include factors that enhance ability to achieve the project objectives (strengths), as well as aspects of the organisation which would inhibit the project (weaknesses). Specific strengths and weaknesses are identified relating to the particular project under consideration, usually limiting this to two or three key strengths and two or three key weaknesses. Example strengths might include process understanding, technical expertise, or team experience. Weaknesses could be poor collaboration management, or tendency to change project priorities.

The workshop then explores the modifying influence of strengths and weaknesses on the specific opportunities and threats associated with each option. This identifies:

Strengths that make a specific opportunity easier to exploit
Strengths that counter exposure to a specific threat
Weaknesses that make it harder to exploit a specific opportunity
Weaknesses that increase exposure to a specific threat

These influences are shown on the SWOT Worksheet. Where a strength or weakness has a modifying effect on an opportunity or threat, the score is changed by $+/-2$ (i.e. a strength adds 2 to an opportunity or threat score, and a weakness subtracts 2). Note that an opportunity or threat scoring zero cannot be influenced by a strength or weakness. Also it is not possible for opportunities to be converted into threats or vice versa, so the modifying influence is limited to zero. Total modified scores are calculated for opportunities and threats, as shown on the example SWOT Worksheet in Figure 6.

A diagram is then constructed (see example in Figure 7), with the vertical axis showing Opportunity Score, and Threat Score on the horizontal axis. The project is usually plotted as a circle, and the diameter of the circle can be used to represent other project characteristics, such as value or Internal Rate of Return (IRR). This allows alternative options or projects to be compared, which can be used for a variety of purposes including tender evaluation, development option feasibility studies, or portfolio assessment.

The risk exposure of the project is indicated by its position on the diagram, with the most favorable position having a high Opportunity Score and low Threat Score. Where this position does not currently exist, actions should be developed to enhance opportunities and/or reduce threats, as shown in Table 1.

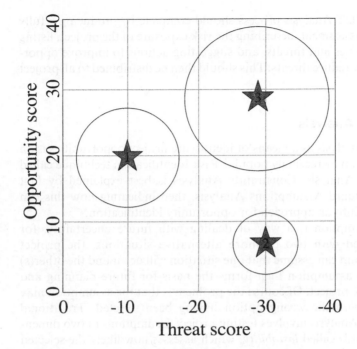

Figure 7 Example SWOT Analysis diagram.

Table 1 Data, Assessment, and Actions for Projects in Figure 7

Project/ option	OPP Score	Threat Score	IRR Score	Assessment	Action
1	20	−10	15	Good opportunities with low threats but limited IRR	Consider whether IRR and/or opportunities can be improved
2	7	−30	10	Low opportunities and high threats, least attractive option for IRR	Reject
3	28	−28	20	Most attractive IRR option with high opportunities but also high threats	Consider whether threats can be reduced

The SWOT Analysis process should complete by producing a fully documented assessment, describing the risk exposure of the project, listing key opportunities and threats, and suggesting actions to improve opportunities and/or reduce threats. This should then be distributed to all project stakeholders.

Constraints Analysis

Constraints Analysis as a means of identifying project opportunities is an adaptation of an increasingly popular Risk Identification technique called Assumptions Analysis. Constraints Analysis is best explained by first outlining standard Assumptions Analysis, then indicating how this can be developed into an approach for opportunity identification.

An assumption is a way of dealing with future uncertainty; for example, faced with two or more alternative situations, the project manager or team can assume that one situation will occur and the other(s) will not. This assumption then forms the basis for future planning and strategy on the project. Of course the problem is that the assumption may prove false, with the wrong option having been selected. Traditional Assumptions Analysis involves testing project assumptions in two dimensions. The first is called *instability*, which assesses how likely the selected assumption is to prove false. Second, the *sensitivity* of the project to each assumption is assessed, determining the significance of any potential effect on the project should the assumption in fact turn out to have been wrong.

Using Assumptions Analysis as a Risk Identification technique first requires project assumptions to have been identified and recorded. This is often done during project initiation, with an assumptions list being included in the project charter. If, however, no list currently exists, this needs to be created as the first step, taking care to explore implicit or hidden assumptions as well as the more obvious explicit assumptions.

After assumptions have been listed, the next step is to test them in the two dimensions of instability and sensitivity. For each listed assumption, the project manager asks two questions: "How likely is this assumption to prove false" (instability), and "If it were false, what impact would that have on project objectives?" (sensitivity). Each dimension is usually assessed qualitatively, i.e., using terms such as High/Medium/Low. Where an assumption is assessed as having high instability and high sensitivity, the project is clearly at risk, since the assumption is likely to prove false and this could have a significant impact on the project. Other combinations such as medium instability/high sensitivity or high instability/medium sensitivity should also be considered as candidate risks. The assessment process can be represented using a two-dimensional grid, as shown in Figure 8.

An assumption can be simply converted into a risk statement by writing it in the form "If the assumption were to prove false (instability), the effect on project objectives would be ... (sensitivity)". This statement corresponds directly to the two dimensions of a risk, with the likelihood of the assumption proving false being equivalent to risk probability, and the effect on objectives describing risk impact.

Assumptions Analysis is proving popular as a Risk Identification technique because it is quick to implement, and it produces risks that are project-specific and linked to project objectives. The quality of the output does, however, depend on the quality of the original assumptions list used as input; if assumptions remain hidden then associated risks will not be identified by this technique. Attention is therefore required to the first step in the Assumptions Analysis process, namely listing assumptions. The

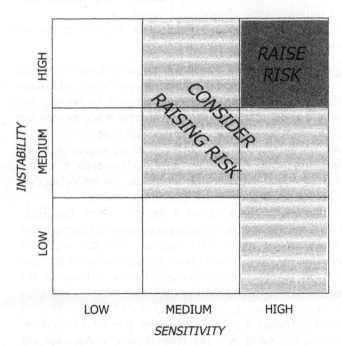

Figure 8   Assessing assumptions and constraints as possible risks.

project team may need assistance in exploring and exposing implicit assumptions, and this should draw on the perspectives of as many stakeholders as possible, who may themselves be making assumptions about the project but without the knowledge of the project manager and team. A facilitated workshop session may be useful to elicit a more complete list of assumptions prior to proceeding with the analysis part of the process.

There is another important shortfall of Assumptions Analysis as a Risk Identification technique—it tends to identify only threats. This is because project managers or team members usually make positive assumptions that things will work out well. Typical assumptions might include:

> All the resources and facilities required for this project will be made available when needed.
> Suppliers and subcontractors will deliver on time and to the agreed specification.
> Senior management (including the project sponsor) will provide a supportive environment for the project.
> Test hardware will be installed and available before system integration testing is due to begin.
> The project manager will remain assigned to this project for its duration.

Each of these is a positive statement that something uncertain will in fact turn out well. As a result, Assumptions Analysis typically results in threat-type risks that the positive assumptions might in fact be wrong, for example that required resources might *not* be available when needed, or that suppliers and subcontractors might *not* deliver on time.

Consequently, the standard Assumptions Analysis approach is not usually appropriate for identifying opportunities. However, a simple modification to this approach can allow it to be used for opportunities, producing a parallel technique known as Constraints Analysis.

In simple terms, assumptions record a decision about the likely positive outcome of a future uncertainty, whereas constraints define limits within which the project must operate. It is common for assumptions to be optimistic ("assume the best case"), and Assumptions Analysis tests these as potential risks, since a false assumption could pose a threat to the project. This approach can be extended to test whether stated constraints might be relaxed, in which case an opportunity might be identified to facilitate achievement of project objectives or enhance project deliverables. The process of Constraints Analysis is therefore similar to Assumptions Analysis, namely to list the constraints, then test the likelihood that the constraint might be false or unnecessary, followed by an assessment of the implications for the project if the constraint were to be relaxed or removed.

As in the case of Assumptions Analysis, the first step in this process is often the most difficult, since many constraints are not explicitly stated. The project team and stakeholders may need to work together, perhaps in a facilitated workshop, to expose and record all relevant constraints before these can be assessed. Examples of constraints might include:

Team size will be limited to six engineers.

This project is assessed as Priority C, and projects of Priority A and B will have preferential access to resources.

Detailed design work cannot start until all high-level design has been approved by the client.

Monthly progress reports to senior management are required.

Documentation will be produced by the in-house reprographics department.

After identifying and listing constraints, the Constraints Analysis process continues with the same two steps of Assumptions Analysis, first testing *instability*, in this case questioning the constraint by asking "How likely is it that this constraint could this be relaxed or removed?" and then testing *sensitivity* by asking "What would be the effect if the constraint was not present?" Constraints with high instability and high sensitivity are candidate opportunities, since there is a good chance that the constraint can be relaxed or removed, and if this is done then there could be a significant positive impact on the project's ability to meet its objectives. Again, the same diagram used for Assumptions Analysis (Fig. 8) can be used to determine which constraints might be most risky.

In the same way that a positive assumption can be tested to reveal a possible threat (if the assumption proved false it would have a negative effect on the project), a constraint can also be tested to expose a possible opportunity (if the constraint could be removed it would have a positive effect on the project). The simple technique of Constraints Analysis therefore offers the project a rapid project-specific means of identifying possible opportunities for further assessment.

Force Field Analysis

Force Field Analysis is a technique developed by the founder of modern social psychology Kurt Lewin (1890–1947) that uses a creative process to identify the *restraining and driving forces* of a change situation. *Restraining forces* hinder change and keep the situation at its current level, whereas *driving forces* initiate and sustain change. This technique is widely used in change management and strategic decision making to identify positive and negative influences on achievement of objectives. It is simple to adopt and adapt this approach to the identification of project risks, by determining those factors that would oppose project success (threats) as well as those that would facilitate it (opportunities).

The basic Force Field Analysis technique has the following steps.

Define and record the situation under study in a simple succinct statement.

List all forces or influences that support or drive change to the situation and those forces or influences that oppose or restrain change.

Rate the strength of each force, using a scale from 1 (weak) to 5 (strong), and show this diagrammatically with the size of each element reflecting its strength.

Assess the current situation to determine the balance of existing forces.

Determine options for action, to encourage change in the required direction. This includes reducing or removing restraining forces, and increasing driving forces.

Of course, many situations are currently stable, and a project is initiated to create change. This means that the organization has three separate goals to accomplish:

Unfreeze—destabilize the current equilibrium in a controlled way, to allow the change to occur. This often occurs outside the scope of the project, but is an essential prerequisite for change.

Change—introduce a deliberate imbalance into the forces to enable change to take place, by increasing drivers, reducing restraints, or both. This is the purpose of the project itself.

Refreeze—once the change is complete, bring the forces back into equilibrium in order to maintain the change. Again this may be seen as part of the project, or may be an organisational activity after the project has completed.

Using the Force Field Analysis technique for Risk Identification requires only a few simple modifications to the standard approach. The "situation under study" is defined as one or more specific project objectives. It is common to treat each project objective separately, conducting a Force Field Analysis for each one, but it is equally possible to perform a single Force Field Analysis for the whole project.

After the project objective(s) have been defined, factors that support their achievement are listed (driving forces), together with those which hinder (restraining forces). This may be done through a brainstorm or workshop session involving the project team and key stakeholders. Each influencing factor is then scored and the force field analysis diagram is produced (see Fig. 9 for an example). This describes the current situation faced by the project: one or more objectives to be achieved are defined, together with factors that are helping and hindering. The status quo usually reflects an equilibrium between the driving forces and the restraining forces, which the project must address in order to deliver the required change. This is reflected in Figure 9, where the scores on each side are approximately equal.

With the situation described in terms of the balance of forces, the diagram allows threats and opportunities to be identified directly. In gen-

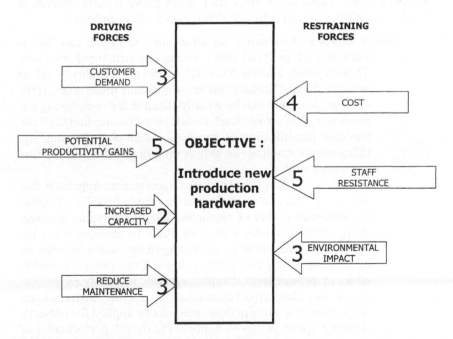

Figure 9 Example Force Field Analysis diagram.

eral terms, there is a threat that each restraining force may become stronger to the extent that it becomes more difficult or even impossible to achieve one or more project objectives. There are also threats that driving forces may be weakened or removed. Similarly opportunities exist to strengthen drivers and remove restraints. The objective is clearly to avoid or minimize threats while exploiting and maximizing opportunities (as discussed in Chapter 7), and the Force Field Analysis will expose areas that should be considered for priority attention, providing an initial assessment of their relative importance in terms of potential effect on project objectives.

Other Approaches

In addition to the three specific techniques outlined above, many other existing approaches might be adopted and adapted to assist the project team in identifying opportunities. The principle is simply to ensure that both types of uncertainty are considered, including those uncertainties able to affect project objectives adversely, as well as those that could have a beneficial effect. Three other areas that might prove fruitful sources of suitable techniques for opportunity identification are as follows:

   Futures thinking: Alternative scenarios for the future can be developed and analyzed using a range of structured methods. These typically address potential upsides and benefits as well as possible threats. Scenario planning is usually undertaken at the strategic level, but can be equally effective for considering microscenarios at project level. Developing "futures literacy" can facilitate identification of beneficial scenarios containing significant opportunities, as well as finding possible futures that are less attractive.

   Value management: This is a structured management approach that aims to optimize value, defined as maximum functions/benefits for minimum outlay of resources. The typical value management process includes a creative phase to generate ideas for improving value, either by increasing functionality/benefits, or by reducing outlay/resources, or both. If this process is undertaken at project level, "value" equates to satisfying project objectives within project constraints. As a result, the techniques and methods of value management can be applied for projects, exposing opportunities to improve the overall performance of the project.

Hazard and safety assessment: A number of well-established techniques exist for analyzing hazards and their potential effect on human safety. These include FMEA (Failure Modes and Effects Analysis), a refined form of causal analysis, and the related technique of FMECA (Failure Modes Effects and Criticality Analysis). Clearly these techniques are designed to expose threats that might lead to failures, but they can also be used in a positive way to analyze opportunities that might produce benefits. Other similar approaches include Hazard and Operability Studies (HAZOPS), which considers causes, consequences, safeguards, and actions; Hazard Analysis (HAZAN) using key words to identify and examine alternatives; and Hazard Analysis with Critical Control Points (HACCP). All of these can be turned to consider the positive dimension, seeking potential upsides that could benefit the project.

It will hopefully be clear from the preceding discussion that many supplementary techniques can be added to the project manager's armory to enhance the ability to identify opportunities. None of these will be effective, however, unless the project manager, the project team, and other key stakeholders share a clear and common understanding that risk includes opportunity, and unless there is also a shared commitment to identify and manage opportunity proactively so as to maximize the project's likelihood of meeting its objectives.

START AS YOU MEAN TO GO ON

Some risk practitioners believe that Risk Identification is the most important phase of the risk management process, since an unidentified risk cannot be managed. Not identifying a risk does not mean it does not exist; it merely means that the project is taking the risk without being aware of it. In the case of threats this will occasionally lead to problems occurring that could have been avoided or reduced if the risk had been foreseen in time to allow proactive action to be taken. In the same way, failing to identify opportunities in advance will result in missed benefits that could have been captured by the project.

This realization of the importance of Risk Identification has rightly led to development of a wide range of powerful techniques in current use, all aiming to assist the project manager and team to spot as many risks as possible, with sufficient lead time to allow them to take appropriate action. Unfortunately the limitations of some of these techniques, together with the preconceptions introduced by force of habit, result in most Risk Identification methods focusing almost exclusively on threats. Where the organization wishes to manage opportunities proactively alongside threats as part of the same risk process, some simple modifications to the existing Risk Identification techniques are required. In addition, there are other techniques that can be used to identify opportunities explicitly, providing the project manager with a more complete view of the totality of risk exposure faced by the project, both upside and downside.

Having completed the Risk Identification phase using the techniques discussed in this chapter, the project manager can move on to the rest of the risk management process knowing that both threats and opportunities have been exposed and can be addressed. If Risk Identification is limited to threats only, there can be no hope of managing opportunities later in the process. But with a list of identified threats and opportunities, the chances of achieving project objectives can be maximized, as threats are reduced or removed and opportunities are increased or captured.

REFERENCES

Amara, R., Lipinski, A. J. (1983). *Business Planning for an Uncertain Future: Scenarios and Strategies.* Oxford: Pergamon Press.

Bartlett, J. (2002). *Managing Risk for Projects and Programmes: A Risk Handbook.* Hook, Hampshire, UK: Project Manager Today Publications.

Bartlett, J. (2002). Using risk concept maps in a project or programme. In: *Proceedings of the 5th European Project Management Conference* (PMI Europe 2002), presented in Cannes France, 19–20 June 2002.

Barton, R. (2000). Soft value management methodology for use in project initiation—a learning journey. *J. Construction Res.* 2(1):109–122.

British Standard BS EN 12973:2000. (2000). *Value Management.* London: British Standards Institute.

Carr, M. J., Konda, S. L., Monarch, I., Ulrich, F. C., Walker, C. F. (1993). *Taxonomy-based Risk Identification.* Technical Report CMU/SEI-93-TR-6, ESC-TR-93-183. Pittsburgh, PA: Software Engineering Institute, Carnegie Mellon University.

Chapman, R. J. (1998). The effectiveness of working group risk identification and assessment techniques. *Int. J. Project Management* 16(6):333–343.

Clark, C. (1989). *Brainstorming: How to Create Successful Ideas.* North Hollywood, CA: Wilshire Book Co.

Clayton, S., Lloyd, B. (2002). Planning for the future. *Professional Manager* 11(4): 32–33.

Coyle, R. G. (1997). The nature and value of futures studies. *Future* 29(1):77–93.

Coyle, R. G. (1998). Futures thinking for decision and risk analysis. *Int J Project Business Risk Management* 2(3):241–255.

Coyle, R. G., Crawshay, R., Sutton, L. (1994). Futures assessment by field anomaly relaxation. *Future* 26(1):25–43.

De Bono, E. (2000). *Six Thinking Hats*. Rev. ed. London: Penguin Books.

Godet, M., Roubelat, F. (1996). Creating the future: the use and misuse of scenarios. *Long Range Planning* 29(2):164–171.

Hall, D. C., Hulett, D. T. (2002). Universal risk project—final report. Available from PMI Risk SIG website *www.risksig.com/articles/UR%20Project%20Report.doc*.

Hillson, D. A. (2002). Structuring a breakdown: the risk management RBS. *Project* 15(7):12–14.

Hillson, D. A. (2002). The risk breakdown structure (RBS) as an aid to effective risk management. In: *Proceedings of the 5th European Project Management Conference* (PMI Europe 2002), presented in Cannes, France, 19–20 June 2002.

Hillson, D. A. (2002). Using the risk breakdown structure (RBS) to understand risks. In: *Proceedings of the 33rd Annual Project Management Institute Seminars, Symposium* (PMI 2002), presented in San Antonio, USA, 7–8 October 2002.

Hillson, D. A. (2000). Project risks—identifying causes, risks and effects. *PM Network* 14(9):48–51.

HM Treasury (1996). *CUP Guidance Note No. 54 Value Management*. London: HM Treasury.

Kaufman, J. (1998). *Value Management: Creating Competitive Advantage*. Los Altos, CA: Crisp Publications Inc.

Kletz, T. A. (1999). *Hazop and Hazan: Identifying and Assessing Process Industry Hazards*. 4th ed. London: Taylor, Francis.

Lewin, K. (1951). *Field Theory in Social Science*. New York: Harper, Row.

McDermott, R. E., Mikulak, R. J., Beauregard, M. R., Mikylak, R. (1996). *The Basics of FMEA*. Productivity Inc.

Miles, L. D. (1989). *Techniques of Value Analysis and Engineering*. 3rd ed. Washington, DC: Lawrence D Miles Value Foundation.

National Advisory Committee on Microbiological Criteria for Foods. (1997). *Hazard Analysis and Critical Control Point (HACCP): Principles and Application Guidelines*. Washington, DC: NACMCF.

Osborne, A. F. (1963). *Applied Imagination: Principles and Procedures of Creative Problem Solving*. 3rd ed. New York: Charles Scribners.

Project Management Institute. (2001). *Practice Standard for Work Breakdown Structures*. Philadelphia: Project Management Institute.

Redmill, F., Chudleigh, M., Catmur, J. (1999). *System Safety: HAZOP and Software HAZOP*. New York: John Wiley.

Ringland, G. (1998). *Scenario Planning: Managing for the Future*. Chichester, UK: John Wiley.

Robinson, R. M., Anderson, K. J., Meiers, S. (2002). *Risk, Reliability—An Introductory Text*. 4th ed. Melbourne, Australia: R2A Pty Ltd.

Schultz, W. L. (1995). *Futures Fluency: Explorations in Vision, Leadership, and Creativity*. PhD thesis, University of Hawaii at Manoa.

Schultz, W. L. (2003). *Futures Studies: An Overview of Basic Concepts*. Available online at *www.infinitefutures.com/essays*.

Slaughter, R. A. (2002). Foresight in a social context. Presented at the Probing the Future: Developing Organisational Foresight. In the Knowledge Economy conference, Strathclyde, Scotland, 11–13 July 2002.

Stamatis, D. H. (1995). *Failure Mode and Effect Analysis: FMEA from Theory to Execution*. USA: American Society for Quality.

Swann, C. D., Preston, M. L. (1995). Twenty-five years of HAZOPs. *J. Loss Prevention Process Indust.* 8(6):349–353.

Thiry, M. (1997). *Value Management Practice*. Philadelphia: Project Management Institute.

Thiry, M. (1998). The benefits of value management throughout the project lifecycle. *Project* 10(9):12–15.

Thiry, M. (2001). Sensemaking in value management practice. *Int. J. Project Management* 19(2):71–77.

Thiry, M. (2001). The new European value management standard: how will it change the way we practice VM? In: *Proceedings of the 4th European Project Management Conference* (PMI Europe 2001), presented in London, UK, 6–7 June 2001.

Thiry, M. (2002). Combining value and project management into an effective programme management model. *Int. J. Project Management* 20(3):221–227.

Thomas, J. (1985). Force field analysis: a new way to evaluate your strategy. *Long Range Planning* 18(6):54–59.

Walsh, B. (2002). Risk diagnosis methodology in Unilever—a case study. In: *Proceedings of the 5th European Project Management Conference* (PMI Europe 2002), presented in Cannes, France, 19–20 June 2002.

5

Qualitative Assessment

Risk Identification identifies risks, or at least it should if done properly. Following the guidelines and approaches outlined in the previous chapter should ensure that all foreseeable risks are listed, representing any uncertain event or set of circumstances that, if it occurs, would have a positive or negative effect on achievement of project objectives. The risks identified should cover all aspects of the project and reflect all stakeholder perspectives.

One might perhaps think that having identified the risks the next step is to determine how best to respond to them. Indeed this may be appropriate for simple projects where there are few identified risks, and on some smaller projects response development can be undertaken at the same time as Risk Identification. However, for most projects the Risk Identification step exposes a significant number of risks, which cannot all be tackled at the same time. It is therefore necessary to prioritize the list of identified risks in some way. This will enable management attention to be focused on the more important individual risks and also on those parts of the project most at risk. Also, where risk management resources are limited, prioritization enables scarce resources to be applied most effectively.

The problem is in moving from the list of identified risks to an understanding of which risks are most "important," and determining which areas of the project are most risky. The "importance" of each individual risk can be assessed either relative to other identified risks to produce a prioritized list, or in absolute terms against some risk threshold that describes the amount of risk that stakeholders deem acceptable. Similarly, the different parts of the project can be ranked by relative riskiness, or the level of risk exposure can be compared against some accepted threshold based on the overall risk appetite of stakeholders.

Clearly, one cannot rely simply on gut feel or intuition to pick out which risks to focus on first, or which parts of the project require priority risk management action. At best this could only be subjective, depending on the perspective and preferences of the person making the judgment. Some structured approach to risk prioritization is therefore required, which assesses all risks within the same objective framework, and which is not dependent on the person making the assessment. This is the aim of the Risk Assessment phase of the risk management process. This chapter discusses the traditional approach to Risk Assessment, indicating how to prioritize opportunities alongside threats.

QUALITATIVE OR QUANTITATIVE?

It should be remembered at this point that there are two fundamentally different approaches to assessing the importance of risks and the riskiness of various parts of the project. The typical risk management process outlined in Chapter 2 (and illustrated in Fig. 2 of that chapter) includes a step called "Risk Assessment," but recognizes that this can be undertaken qualitatively or quantitatively. Qualitative Risk Assessment aims to describe each risk using words and phrases, with the aim of enabling the risk to be understood in sufficient detail to allow development of appropriate responses. The qualitative process also groups risks to determine patterns of risk exposure across the project. Quantitative Risk Analysis, on the other hand, is about using numbers to represent the dimensions of each risk, then performing some statistical or numerical analysis to determine the overall effect of risks acting together on project objectives. This chapter deals with qualitative Risk Assessment and how it can be used to address upside opportunities as well as threats, while quantitative Risk Analysis is covered in Chapter 6.

The relationship between qualitative Risk Assessment and quantitative Risk Analysis should be clarified before going on to describe these phases. Most risk practitioners agree that qualitative Risk Assessment is not optional for the majority of projects, since it is vital for effective risk management that each risk should be properly understood. By contrast, quantitative Risk Analysis is viewed by many as an additional process that may be implemented for a particular project if the circumstances demand it, but that is not required as part of the risk management process for all projects. Where qualitative Risk Assessment is mandatory for all projects, quantitative Risk Analysis may be seen as an optional extra. This will be explored further in Chapter 6.

DIMENSIONS OF ASSESSMENT

The traditional approach to qualitative Risk Assessment is based on the understanding that risk has two dimensions. The definition of risk derived earlier (discussed in Chapter 1 and summarized in Fig. 5 of that chapter) is *"any uncertainty that, if it occurs, would affect one or more objectives."* This includes one dimension related to uncertainty, and another covering the effect on objectives (noting that the effect can be either negative, for threat-type risks, or positive, for opportunities). To determine the "size" of any given risk, it is necessary to evaluate both of these two dimensions, and then use that evaluation to prioritize risks.

This two-dimensionality of risk was first stated by the French theologian and philosopher Antoine Arnauld (1612–1694), who wrote in 1668, "Fear of harm ought to be proportional not merely to the gravity of harm, but also to the probability of the event." Arnauld's friend Blaise Pascal (1623–1662), who developed the theory of probability along with Pierre de Fermat (1601–1665), made a similar comment but related it to upside risk, when he said, "The excitement that a gambler feels when making a bet is equal to the amount he might win times the probability of winning it." These two quotes demonstrate that the idea of treating both threat and opportunity similarly is not new, since it was being considered by the earliest thinkers in the field of risk management.

The Uncertainty Dimension

The degree of uncertainty associated with a risk can be described using the term "probability" to reflect how often one might expect the risk to occur. A purist mathematician would rightly argue that this is not a strictly accurate use of the term, and might prefer to use "frequency," "chance," or "likelihood," but "probability" has become established in risk terminology as the recognized descriptor for the uncertainty dimension.

Probability can be expressed in two basic ways: either using descriptive labels such as High, Medium, or Low; or numerically with a value between zero and one, or between 0 and 100%. There are difficulties associated with both methods of describing probability:

Problems Associated with Descriptive Labels

Labels are ambiguous and are therefore open to subjective individual interpretation, with different people understanding them in different ways. For example, one person might consider that the statement "There is a low probability of rain today" means that rain is extremely unlikely, perhaps corresponding to a 1:100 chance, whereas another might think that "low probability" means less than 50:50, perhaps one in three or one in four. Similarly the phrase "high probability" might suggest a 99% chance of occurrence to one person, while for another it could mean anything above 50:50. As a result it is necessary to define and agree what is meant by the labels if they are to be used to describe the probability of risks occurring on a project. Another issue to be addressed is the degree of granularity required for assessing this uncertainty dimension of risks. For example, three levels might be sufficient (High, Medium, and Low), or it may be considered necessary to use four, five, or six levels (extending with Very High, Very Low, Extremely Low, etc). The number of levels should be driven by the complexity of the risk process, as determined in the Definition phase (see Chapter 3).

Problems Associated with Numerical Values

The alternative approach of using numbers to describe the uncertainty dimension overcomes the problem of differing interpretation, since "53% probable" is unambiguous. There are, however, several difficulties here, with the first relating to precision. A number is precise, but it is being used here to describe uncertainty, which is inherently imprecise. So within the limits of knowledge and estimating precision, probability assessments of 53% and 46% may be essentially equivalent, since both mean "about 50:50." The second problem with using precise numbers to describe probability is the data source. Since many project risks are unique to the project, there is often little or no relevant historical data on how often a similar risk might have occurred in related projects. This means it is not usually possible to determine the probability of any particular risk with any degree of precision based on previous experience. Finally, despite the unambiguous nature of numbers, it is still possible for people to interpret them in different ways. For example, does the forecast of "20% chance of rain in southern England today" mean that it will definitely rain for the whole day across 20% of the area, or that all of southern England will experience rain for 20% of the day, or if you make five journeys you will get rained on once, or that it is most likely not to rain at all and if it does you are just unlucky? These difficulties mean that numbers should only be used to describe probability where there is previous relevant historical data, ranges should be used in preference to single numbers, and clear definition is essential.

Solutions

The first difficulty of ambiguous interpretation can be overcome by clearly defining what is meant by each probability label. This can be done at a number of levels of detail. For example, the labels can be expanded into descriptive phrases that give a better indication of their intended use, as follows:

Probability label	Phrase
Very Low	Improbable
Low	Unlikely
Medium	Possible
High	Probable
Very High	Likely

While this is better than using simple labels, there is still considerable room for interpretation, allowing different individuals to overlap between the phrases—one person's "unlikely" could be another's "improbable." Empirical research on differing individual interpretations of probability-related terms is summarized and compared in Table 1 to illustrate this issue, showing the wide variation in what these phrases might mean to different people. A further refinement over using descriptive phrases is to give indicative numerical probabilities for each label, as follows:

Probability label	Indicative probability
Very Low	5%
Low	25%
Medium	50%
High	75%
Very High	90%

Table 1 Interpretation of Probability-Related Terms

Descriptive phrase	Hamm (1991)	Lichtenstein & Newman (1987)	Boehm (1989)
Rare	5%	1–30%	—
Very/highly unlikely	10%	1–30%	0–10%
Seldom	15%	1–45%	—
Improbable	—	1–40%	1–45%
Not very probable/ probably not	20%	1–60%	1–40%
Unlikely	—	1–45%	1–45%
Fairly unlikely	25%	1–75%	—
Less than half the time	45%	5–50%	—
Probable	—	1–99%	55–85%
Better than even	60%	45–90%	50–60%
Likely	—	25–99%	55–85%
(Very) good chance	75%	25–95%	60–85%
Quite likely	80%	30–99%	—
Very probable	85%	60–99%	—
Highly probable	90%	60–99%	—
Almost certain	95%	—	80–99%

Instead of using percentages, an alternative representation could be to use odds, as follows:

Probability label	Odds
Very Low	1:20
Low	1:4
Medium	1:2
High	3:4
Very High	9:10

While this may be an improvement over the use of phrases, this approach suffers from the second set of difficulties mentioned above, namely spurious precision and data sources. If each label is defined merely in terms of a single number, problems will arise for risks whose probabilities are assessed as being between the given values (e.g., whether a risk with a 33%

probability of occurrence should be rated Low or Medium), and with lack of good data sources to estimate the probability at the required level of precision.

A solution commonly adopted is to combine descriptive probability labels with numerical definitions, in other words using labels such as High, Medium, or Low, but associating each with a percentage range, for example:

Probability label	Range
Very Low	1–10%
Low	11–30%
Medium	31–50%
High	51–70%
Very High	71–99%

This defines the labels in an unambiguous way, but retains the required degree of uncertainty by presenting ranges rather than precise single figures. There can, of course, still be disagreements over the boundaries. For example, a risk whose probability is below 50% could be just a little below (i.e., Medium probability), or some way below (say Low probability), or a long way below (with Very Low probability). Nevertheless the provision of a range for each label seems to offer the best of both worlds, with the precision of numbers combined with the uncertainty of a range.

The set of percentage ranges given in the last table above is just one of many possible alternatives, and different sets are clearly possible (for example Very Low = < 5%, Low = 5–10%, Medium = 11–25%, High = 26–50%, Very High = > 50%). The key point is not the specific detail of the particular set of definitions chosen, but that there should be an agreed language for describing the uncertainty dimension of risks, allowing this to be assessed in a consistent manner.

It should be noted that the set of probability ranges given in the final example table does not include zero or 100% as the two extremes. This is because the range is to be used for assessing risks, which are by definition uncertain. Zero probability represents an impossible event and 100% means it is certain to occur, and neither of these two circumstances involves uncertainty. Consequently definitions of probability ranges should start above zero and end below 100%, although the precise start and end points to be used in a given project are a matter of judgment.

When considering the two types of risk (threats and opportunities), it is clear that both have a similar dimension of uncertainty, since neither is a definite event. As a result, a common approach can be applied to assessment of the probability of occurrence of threats and opportunities, using the same defined scale of probability to describe how likely any particular threat or opportunity is to occur.

Effect on Objectives

The second dimension of any risk is the effect it would have on one or more project objectives if the risk actually occurred. This is usually called "impact," although other words such as "consequence" or "effect" can be used. The impact dimension can also be described using labels or numbers in the same way as described above for probability, with the same shortcomings and caveats.

Again there are several ways of describing impacts, with increasing levels of sophistication. For example descriptive phrases can be used, or percentage effects on project duration or cost, or specific values might be given, as shown in Table 2.

The main difference between probability and impact is that each risk has only one probability, representing an assessment of how uncertain it is. A single risk could, however, have more than one impact, since it can affect one or more project objectives. For example, there might be a threat risk on a software development project that previously unknown errors might be exposed during final system acceptance. If this occurs, there would be impacts on both project timescale and the quality of the final deliverable. Another opportunity risk might be identified that a subcontractor might be taken over by a more efficient competitor, leading to possible reduced

Table 2 Alternative Definitions of Impacts

Impact Label	Phrase	% effect	Specific values
Very Low	Insignificant	1–2% change	Up to 1 week, or up to $1K
Low	Minor	3–5% change	1–2 weeks, or $1–5K
Medium	Significant	6–10% change	3–4 weeks, or $6–10K
High	Major	11–20% change	5–8 weeks, or $11–25K
Very High	Critical	>20% change	Over 8 weeks, or >$25K

delivery times and decreased costs. As a result, it is necessary to assess the impacts of identified risks against each project objective. This usually includes project time and cost, but projects do, of course, have other objectives that depend on the project type, such as quality, performance, technical functionality, reliability, operability, regulatory compliance, safety, security, company reputation, environmental impact, etc. It is therefore necessary to define impact labels against the various project objectives. Table 3 shows an example for a software development project, where a third project objective is included covering the performance of the delivered system.

The linkage between impacts and project objectives is also important when considering the definition of labels such as High/Medium/Low in relation to impact. A high impact on a small project is likely to have a lower numerical value than a high impact for a large project. For example in Table 3 the term "Low cost impact" is defined as a change in overall project cost of between $100K and $250K. On many small projects the total budget would be considerably less than this, and the definition of "Low cost impact" would be correspondingly lower. Impact definitions must therefore be project-specific, and need to be set by the organization for each project.

This raises the question of how to define impact ranges for a particular project. The aim is to produce a set of definitions that captures the degree to which the impact of a risk is acceptable to an organization. Clearly a range of impacts is possible for risks, with negative impacts for threats, and positive impacts for opportunities. Threats that materialize could result in a major catastrophe that would result in project cancellation or significant damage to the organization, or they may just give rise to a minor inconvenience that can easily be absorbed within the project. Some

Table 3 Typical Project-Specific Scales for Probability and Impacts

| Rank | Prob | Impact on project objectives ($\pm$) | | |
		Time	Cost	Performance
NIL	—	No change	No change	No change in performance
VLO	1–10%	<1 week	< $100K	Effect on 1 minor parameter
LO	11–30%	1–2 week	$100–250K	Effect on >1 minor parameters
MED	31–50%	3–6 weeks	$250–500K	Minor effect on key parameters
HI	51–70%	7–12 weeks	$500–1000K	Major effect on key parameters
VHI	71–99%	>12 weeks	> $1000K	Effect on overall functionality

upside risks are "golden opportunities" that could significantly enhance the value and benefits from the project and could revolutionize the organization, whereas others might just be nice to have if they occurred but would not represent significant additional benefits. The values in the impacts definition table are set so that the highest impact (e.g., Very High) describes either a level of possible negative impact that cannot be accepted or a level of potential upside that must be exploited. The lowest levels (Very Low) represent impacts on project objectives that would not seriously affect the progress or performance of the project. Intervening levels are spread between these two extremes.

The levels of unacceptable threat or unmissable opportunity are driven by the risk attitude and risk appetite of the project stakeholders, and it is not possible to define impact levels without knowing these. However, many projects proceed without a clear understanding of where the threshold of acceptability lies. This is usually detailed during the Definition phase of the risk management process (see Chapter 3), where stakeholders document in the Risk Management Plan the degree of risk they deem acceptable for this project.

Of course, many organizations restrict their risk management process to dealing with threats, as discussed earlier in Part I. Under this approach impact definitions are exclusively negative, representing time delays, additional cost, performance shortfalls, etc. Those organizations wishing to take a broader approach that deals with both threats and opportunities together within a common risk process will, however, need impact definitions covering both upside and downside. A single set of impact definitions can be used such as those in Tables 2 or 3, where the definitions of impact are interpreted either negatively for threats or positively for opportunities. For example, using Table 3, a threat with a High time impact might delay the project by 7–12 weeks if it occurred, whereas an opportunity whose cost impact was Medium might result in a cost reduction in the range $250–500K. The same table of impact definitions is used, but the sign depends on the type of risk—negative impacts for threats and positive for opportunities. An alternative would be to use two tables of impact definitions, one for threats and another for opportunities, defining different ranges for each.

In one sense the actual definitions used do not matter. The important thing is to define scales prior to undertaking the assessment phase of the risk management process, to provide a common framework against which individual risks can be assessed. Scales should be set by the stakeholders for each project and communicated to all participants in the risk process, so that each risk is assessed consistently, allowing them to be prioritized relative to each other.

Using Two Dimensions to Assess Risk

It is an interesting exercise to ask people to identify their "biggest current risk" and then to explore how they defined and measured "biggest." The definition of risk as "any uncertainty that, if it occurs, would affect one or more objectives" includes two dimensions: uncertainty (described as "probability") and effect on objectives (described as "impact"). The assessment of "biggest" is often made without explicit reference to these two dimensions, with individuals instead tending to rely on gut feel or some other factor. The most common other factors used when trying to determine the biggest risk include:

Manageability (the biggest risk is the least controllable)
Familiarity (lack of experience, knowledge, or skill increases risk)
Frequency (the risk might arise often in a given period of time)
Temporal proximity (either the risk or its possible impact is nearest in time)
Personal propinquity (the risk that could affect me most)
Corporate vulnerability (the risk that could affect my project or organization most)

Although these other factors are important components of risks, and need to be understood, they are commonly excluded from the formal Risk Assessment process, which relies instead on the two main dimensions of probability and impact. But how is it possible to use two dimensions in assessing risk? Is one more important than the other?

Some people when identifying their "biggest current risk" are driven by probability (the biggest risk is the one most likely to happen), whereas for others impact is more important (the biggest risk is the one that could have the largest effect on the project). Research suggests that more people consider impact than probability when sizing or ranking risks, even when such consideration is subconscious or implicit.

It is, however, not possible to rank risks based only on a single dimension, such as how likely they are to occur. Neither can the potential impact be used in isolation as a basis for prioritization. Both dimensions are important, and some way of combining the two is required for a proper assessment of risk. The most common method is to use a two-dimensional matrix known as a Probability-Impact Grid on which risks are plotted, with the position representing the combination of probability and impact for each risk.

VISUALIZING THE RISK ASSESSMENT FRAMEWORK

The discussion of Risk Assessment so far has concentrated on defining the framework against which identified risks can be assessed. Each risk has two dimensions, known as probability and impact, and these can each be described qualitatively using defined labels, such as High, Medium, or Low. The sections above have outlined ways in which such labels can be defined, allowing each risk to be assessed against a common set of criteria. One probability will be determined for each risk, assessing how likely it is to occur. There may, however, be several different impacts, measuring the effect of the risk on various project objectives if it occurs, including both negative impacts (threats) and positive impacts (opportunities). To prioritize risks for further attention and action, it is now necessary to compare the assessed risks against each other. The best way to do this is with a visual representation of the assessment.

One of the first decisions to make before commencing a qualitative Risk Assessment is the level of granularity to be used for each of the dimensions of probability and impact. It is conventional for both dimensions to have the same number of degrees of detail, though this is not mandatory. The most simple approach would be to use a 2×2 framework, which defines probability as High/Low, and also defines impact as High/Low. This allows identified risks to be divided into four categories:

> High probability/High impact
> Low probability/High impact
> High probability/Low impact
> Low probability/Low impact

These can be shown graphically as a matrix, usually known as a Probability-Impact Grid (or P-I Grid), as in Figure 1(a). Clearly risks with probability and impact both high are top priority (quadrant "1" in the figure), and those with both low are bottom priority (quadrant "4"). Of the other two quadrants, most people agree that "Low probability/High impact" ranks higher than "High probability/Low impact," thinking that a small chance of a major impact is more important than a high chance of something insignificant. This confirms the view that impact is more important than probability, as discussed above.

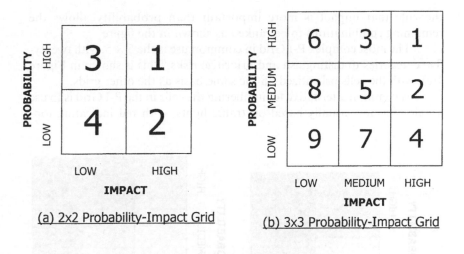

(a) 2x2 Probability-Impact Grid

(b) 3x3 Probability-Impact Grid

(c) 5x5 Probability-Impact Grid

Figure 1 Probability-impact grids.

The 2 × 2 matrix is the least complex approach, but it may not provide sufficient granularity to prioritize risks on a typical project, where there may be a large number of risks that need to be separated into more than four priority groups.

More detail is possible with additional levels in each dimension. A 3 × 3 P-I Grid is shown in Figure 1(b), providing nine categories. Again High/High is the top ranked category, and Low/Low is the bottom. Using the rule that impact is more important than probability allows the remaining combinations to be ranked as shown in the figure.

The most complex P-I Grid in common use is the 5 × 5, with twenty-five categories to distinguish and prioritize risks. This is shown in Figure 1(c), with the cells prioritized on the same basis as the other grids.

A common alternative to numbering the cells in the P-I Grid is to use a color scheme, usually based on traffic lights, with red indicating top-

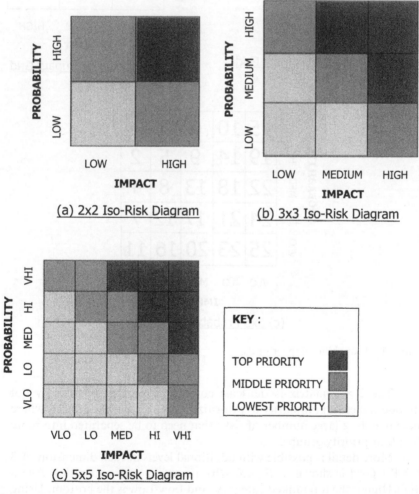

Figure 2 Iso-risk diagrams.

priority risks, amber or yellow for middle-priority risks, and green for lowest-priority risks. This allows a higher level of granularity to be used for assessing risks, but then groups assessed risks into a fewer number of overall priority levels. Figure 2 illustrates this (with red/amber/green shown as dark-gray/mid-gray/light-gray). This approach is used to create an "iso-risk diagram," with zones containing risks of similar priority. The boundaries on the iso-risk diagram can be modified to reflect the risk appetite of stakeholders or the risk threshold for a particular project. Moving boundaries toward the top right of the diagram represents increasing risk appetite or a higher risk threshold, since a smaller "red zone" means that fewer risks will be treated as "top priority" and more risks will become "lowest priority" in the bigger "green zone." Similarly adjusting the iso-risk boundaries toward the lower left would be suitable for a project with a lower risk threshold where more risks are to be treated as top priority.

PERFORMING THE QUALITATIVE ASSESSMENT

Having defined the terminology to be used for the assessment (i.e., project-specific definitions of probability and impacts), and determined the level of granularity of the assessment methodology (i.e., whether to use a P-I Grid that is 2×2, 3×3, or 5×5, for example), the next step is to assess each identified risk, and then prioritize risks for further management attention.

Assessment is a simple matter of taking each risk in turn and comparing it with the table of definitions of probability and impacts, using a framework similar to the one in Table 3. For each risk, it is necessary to assess how likely it is to happen, and what its effect(s) might be on one or more project objectives if it were to occur. This assessment can be done by the project manager alone, or by the project team in a group setting, or in consultation with other stakeholders. Where more than one person is involved in assessing risks, there is often a discussion over the "correct" assessment, e.g., whether the probability is High or Medium, or whether the potential impact might be Low or Very Low. Such a discussion can be very valuable in leading to an improved understanding of the risk, and exposing alternative assumptions held by people with differing perspectives. They should also remember that there is no definitively "correct" assessment, since the future is inherently uncertain—all assessments are best estimates based on currently available information and the perspectives of the assessors. Nevertheless it is important for the people doing the assessment to reach a consensus on each risk, documenting their assumptions for future reference.

Having scored identified risks in the two dimensions of probability and impact, each risk can then be plotted on the P-I Grid in the appropriate position. For example, a risk assessed as having a Low probability of occurrence and High potential impact would be plotted in the Low/High cell. This would correspond to position 2 if a 2 × 2 P-I Grid was being used as shown in Figure 1(a), position 4 in a 3 × 3 grid like Figure 1(b), or position 12 in a 5 × 5 grid such as Figure 1(c).

After all assessed risks have been plotted, they can then be prioritized based on their position in the grid, using either the cell positions, as in Figure 1, or the iso-risk zones, as in Figure 2.

It is important to note that this prioritization process can be used to handle both threats and opportunities equally, if the same approach is used to assess both types of risk. Imagine a project where identified risks include both upside opportunities and downside threats, and where project-specific probability and impacts have been defined as in Table 3. Each threat is assessed against these scales, with the impact definitions being treated as negative (e.g., a Low time impact means a delay of 1–2 weeks). Each opportunity is similarly assessed using the same scales, but interpreting impacts as positive (e.g., a Low time impact means a saving of 1–2 weeks). Both threats and opportunities can then be plotted on the same P-I Grid and prioritized as described above, for example using a 5 × 5 iso-risk diagram as in Figure 2(c). This would then rate as "top priority" those risks appearing in the red zone (dark gray in the figure). In the case of threats, these are uncertain events that have a significant chance of occurring and could cause major damage to the project—clearly these require focused attention. Opportunities appearing in this zone also have a good chance of happening, and if they occur they would significantly benefit the project—these also require attention. Whether risks have upside or downside impacts, those appearing in the "top priority" zone are either show-stopper threats or golden opportunities. In both cases they should be treated seriously, to avoid or minimize the worst threats, and also to exploit or maximize the best opportunities.

EXTENDING THE PROBABILITY-IMPACT GRID

The standard P-I Grid as discussed above presents a means of visualizing the assessment of each identified risk in the two key dimensions of probability of occurrence and impacts against objectives, allowing prior-itization for further management attention and action. There are, however, a number of potential shortfalls in this two-dimensional representation, which can be overcome with some simple modifications. These include dealing with risks with impacts against more than one project objective, representing threats and opportunities, and improving the approach to prioritization.

Handling Multiple-Impact Types

The first aspect to consider is that risks can affect several project objectives, as discussed above. For this reason, the "impact" dimension is usually defined against each project objective separately, as shown for example in Table 3. If this approach is adopted, the assessment of each identified risk would result in one estimated probability value (represent-ing the chance that the risk might occur), and at least one estimate of impact against project objectives. There will certainly be some identified risks that would affect more than one project objective if they occurred. This raises the question of how to show such multiple-impact risks on a P-I Grid. For example, if a risk was assessed as having a Medium probability of occurrence, but which, if it occurred, would have a Low effect on timescale and a High impact on project cost, how should this be shown on a P-I Grid?

A number of alternative solutions exist to the question of how to represent the multi-impact risk on a P-I Grid.

1. One option is to select the highest level of impact and show only that on the P-I Grid, since this represents the most significant position (i.e., "worst-case" for threats, and "best-case" for opportunities). In the case of the risk with Medium probability, Low time impact, and High cost impact, this would be plotted in the Medium/High position, driven by the cost impact and ignoring the time impact. Although this is the most simple solution, it results in loss of information about the risk, since the P-I Grid is not showing the lower impacts. It is, however, an option adopted quite widely, since it allows risks to be prioritized based on their highest impact.

2. A second solution is to use multiple P-I Grids, with one for each impact type. The project would then have a P-I Grid for time impacts, another for cost, and perhaps a third for performance.

In this case, the example risk described above would be plotted on the time grid in the Medium/Low position, and on the cost grid as Medium/High. While this option captures all the available information on a risk, it can appear complicated, with the same risk appearing on several grids, and this in turn can lead to confusion.

3. A third option, which also retains all the assessment information about each impact dimension but which avoids the complexity of multiple P-I Grids, is to use a single grid but to code the impact type. Thus the example risk might appear twice on the P-I Grid, in both the Medium/Low and Medium/High positions, but using the suffix "T" to indicate the position of the time impact, and "C" for the cost impact. If each risk has a unique identifier, for example from a Risk Register or Risk Log, this can be used in this way. Again this option records all the information but can result in a complicated P-I Grid with many entries, especially if there are a large number of identified risks.

4. One might imagine another alternative that plots the "average position" of the risk. The example risk with Medium probability and two impacts (Low time impact and High cost impact) would then be plotted in the Medium/Medium position. Although this appears simple, it is in fact an oversimplification, since vital assessment information is lost. Worse, the resultant prioritiza-tion can be misleading. For example the "average impact" would be the same for a risk with two impacts that are Very Low and Very High, and for another risk with three Medium impacts. However, these two risks should perhaps be treated differently since the first includes the potential for a major impact on one project objective, whereas the second is mediocre in all impact dimensions. As a result, this averaging option is not recom-mended as a solution to plotting the multiple-impact risk.

5. A final option might be to use a multidimensional representa-tion, for example a 3-D P-I Grid (cube) with probability, time impact, and cost impact on each axis. Although this is theoret-ically possible, in practice it would be limited to three dimen-sions, and even these would be hard to read.

This range of possible representations raises the question of which should be used for a particular project. No single preferred solution is right for all projects. The risk process is intended to be scalable, with the level of detail matching the risk challenge of the project, as documented in the Risk Management Plan (see Chapter 3). As a result, one project may decide to use a simple approach to Risk Assessment, perhaps selecting a smaller P-I Grid (say 2 × 2 or 3 × 3) and plotting only the highest impact type. It may, however, be appropriate for another project to use a more detailed approach, with multiple 5 × 5 P-I Grids showing separate impact dimensions.

Representing Both Threats and Opportunities

The fact that identified risks include both threats and opportunities can be treated simply as a particular case of different impact types. In the same way that the standard P-I Grid can be modified to handle different impact dimensions, it can also deal with risks that have negative impacts (threats) and positive impacts (opportunities). Alternatives exist similar to options 2 and 3 above, namely:

Use of a naming convention on a single grid, e.g., "T" or "−" for threats and "O" or " + " for opportunities.
Use of two P-I Grids, one for plotting threats and one for showing opportunities.

While the single-coded grid presents all the information in one place, it can be confusing to mix upside and downside risks, especially if different definition scales have been used for threat and opportunity impacts. As a result, a double P-I Grid is recommended. This can be a simple duplication of the standard P-I Grid, as shown in Figure 3(a), which shows two 5 × 5 grids alongside each other, using one to plot threats and the second for opportunities.

A useful alternative is the so-called "mirror P-I Grid," where the opportunity side is rotated by reversing the impact scale, creating a symmetrical double grid as in Figure 3(b). The advantage of the mirrored version is that the top priority threats and opportunities appear together in a V-shaped zone in the center of the double grid. This so-called "Attention Arrow" covers the worst threats and the best opportunities, which together should be the focus of management attention and action. The size of this zone can be changed to reflect the amount of management effort available for risk management, or in response to the level of acceptable risk on the project. For example, where risk management effort is limited or the risk threshold is high, the Attention Arrow might be quite small, focusing only

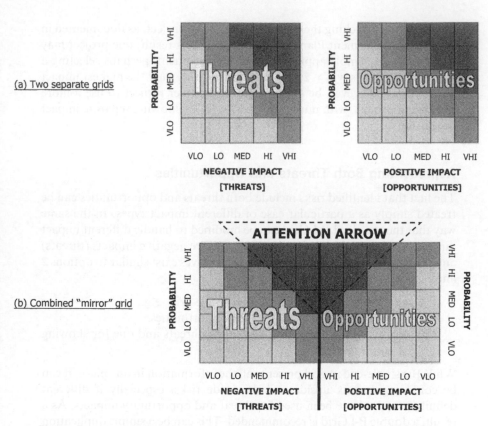

Figure 3 Representing threats and opportunities on P-I Grids.

on the few most significant risks. Where more resources are available for risk management, or the risk threshold is low, a larger Attention Arrow will include more risks for priority action.

Improving Prioritization

The standard P-I Grid allows risks to be categorized into a number of groups for further attention, based on their position in the grid. For example, the iso-risk zoned P-I Grid (as in Fig. 2) divides risks into high, medium, and low groups. Grids with numbered cells as in Figure 1 provide greater resolution, with a larger number of categories than just the three high/medium/low zones.

However, the standard grids described above and illustrated in Figures 1 and 2 are all symmetrical, with zone boundaries lying at 45 degrees across the grid. This does not properly reflect the intuitive perception of most people that impact is more important than probability when assessing the significance of a risk. For example, if Risk A has a low probability of a high impact, and Risk B has a high probability of a low impact, most people would assess Risk A as more important than Risk B. The relative importance of these two cases becomes more clear at the extremes. For example, when considering threats, a small chance of a disaster is more important than near-certainty of a minor inconvenience. Similarly for opportunities, it is more important to try to capture a golden opportunity even if there is only a small chance of succeeding, rather than chasing something that is easier to obtain but would give only minor benefits to the project.

It is possible to skew the zones in the P-I Grid to reflect this required weighting of impact over probability. The simplest way to implement this is by using a scoring system that is linear for probability and nonlinear for impact. An example is shown below.

Scale	Probability score	Impact score
VLO	0.1	0.05
LO	0.3	0.1
MED	0.5	0.2
HI	0.7	0.4
VHI	0.9	0.8

The scoring system is used by assessing for each risk where its probability and impact lie on the scale, then selecting the corresponding scores and multiplying them together. This produces a nondimensional number commonly called a P-I Score. For example, using the scoring scheme in the above table, a risk with Medium probability and High impact would have a P-I Score of 0.5 × 0.4, i.e., 0.20.

Applying this particular scoring system to the 5 × 5 P-I Grid of Figure 2(c) produces a set of P-I Scores as shown in Figure 4, ranging from 0.005 to 0.72. The numbers themselves have no absolute meaning, and cannot be used to determine contingency sums or project duration, but they allow the relative importance of individual risks to be ranked against

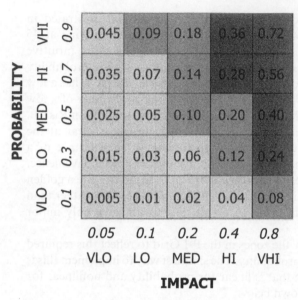

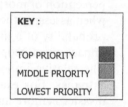

Figure 4   5 × 5 Iso-Risk Diagram based on P-I Scores.

each other on a consistent basis. It is also important to note that the specific numbers in the above table are merely one example of a P-I Scoring scheme, and other schemes are equally possible and valid. For example on a 3 × 3 P-I Grid it is possible to associate scores of 1/2/3 with Low/Medium/High probability, and 1/2/4 for Low/Medium/High impact, creating a range of P-I Scores between 1 and 12.

Having produced a set of P-I Scores that represent both dimensions of probability and impact, the iso-risk zones can then be defined using these numbers. The example in Figure 4 sets the "High risk" red zone of top priority risks where the P-I Score is greater than 0.20. The "Low risk" green zone has scores less than 0.08.

The combination of linear and nonlinear scales results in skewed zones on the P-I Grid, with more emphasis being given to impact. Thus the red zone of top-priority "High risks" includes risks with Low probability and Very High impact, in line with intuitive expectations.

These zones can be moved as discussed earlier to reflect changes in risk appetite or stakeholder risk thresholds, and this is achieved simply by changing the P-I Score associated with zone boundaries. For example, a project that is required to take less risk might move the red and green zone boundaries toward the lower right of the grid by changing the definition of the red zone to include all risks with P-I Score greater than 0.10 (instead of 0.20), and redefining the green zone to cells with score less than 0.04. Moving the zone boundaries in this way results in a bigger red zone, with more risks being treated as top priority, thus ensuring that the project takes less risk. The converse position can also be created for a project with a higher risk threshold that is allowed to take more risk, where a smaller red zone and larger green zone results in fewer High risks—this is achieved by using higher threshold values of P-I Scores for the zones (say green zone < 0.12, and red zone > 0.28).

If a P-I Scoring scheme such as the one outlined above is employed, it allows a more detailed analysis of the risks identified on the project. Such analysis can include the following elements:

Rank individual risks. Clearly it is a simple matter to produce a prioritized list of risks, based on their P-I Scores.

Identify the most important risks. These can be defined as all risks in the red zone, or the "Top ten" risks from the prioritized list (or any other number of top risks).

Assess effectiveness of proposed risk responses. This can be determined by monitoring the change in P-I Score of individual risks before and after responses are implemented. It can also be done prospectively to assess the likely effect of responses before implementation.

Compare different types of risk exposure. P-I Scores can be calculated separately for different impact dimensions on each risk, and the degree of time risk exposure can then be compared to the overall cost risk exposure (using P-I Scores rather than duration or monetary scales).

Analyze overall project risk exposure. P-I Scores can be totaled for the project, indicating the total amount of risk to which the project is exposed, and trend analysis can be performed during the lifetime of the project to determine how risk exposure is changing.

Compare options. The relative risk associated with different project options can be compared in a consistent manner by using a common P-I Scoring scheme to assess risks. For example, an organization may be considering whether to subcontract development of an element of the project or to do the work in-house.

The risks associated with each option can be identified and then assessed using P-I Scores, comparing the total P-I Score for each option to determine the lower-risk alternative. A similar approach can be used to assess competing tenders in the pre-project phase.

When P-I Scores are used for analysis in this way, it is important to distinguish between threats and opportunities, to avoid confusion. A high P-I Score could be associated with either a serious threat or a significant opportunity. However, the desired change as a result of management of each type of risk is entirely different. For threats, the aim is to reduce the risk as indicated by a lower P-I Score, preferably removing the threat and setting the P-I Score to zero. Conversely, the goal for each opportunity is to maximize it as far as possible, increasing the P-I Score as a result, and aiming to make the opportunity happen. Because the desired direction of change is different for threats and opportunities, it is not possible to monitor total P-I Score for the project, combining both. It is also not possible to use positive P-I Scores for opportunities and negative scores for threats, since these would merely cancel out in the total score and conceal the true position. It is therefore necessary to distinguish the two types of risk and monitor trends in their P-I Scores separately.

At first sight the use of P-I Scores appears to be complex, requiring multiple calculations to enable risks to be scored and prioritized. Fortunately, implementation of a P-I Scoring scheme is not difficult, and can be automated using a spreadsheet or database application. Proprietary risk software also commonly incorporates calculation of P-I Scores for risk prioritization, so that the user does not have to perform calculations manually.

CATEGORIZING RISK

The preceding discussion has concentrated on considering and assessing identified risks individually. The probability and impact(s) of each risk are assessed in turn and used to determine the relative significance of the risk, to focus further management attention and action. After risks have been prioritized, the traditional approach is to concentrate on the top risks.

An important part of Risk Assessment, however, is to consider the overall risk exposure of the project, by examining the distribution of identified risks and looking for concentrations of risk. This can be achieved by grouping risks together in various ways. This more structured approach is an important element of Risk Assessment, since risks do not affect the project one at a time; instead they occur in groups. It is therefore vital to consider risk groupings to identify concentrations of risks, both to be aware of the degree of risk faced by the project or the organization, and to enable effective generic responses to be designed that can tackle groups of risks together.

Two types of risk categorization are commonly considered during the Risk Assessment phase, namely causes or sources of project risks, and the areas of the project affected by risks. A number of techniques are presented below to analyze the distribution of identified risks by root causes or common sources, and by project areas affected.

Exploring Root Causes of Risk—Modifying Existing Techniques

To understand which areas of the project might require special attention, and whether there are any recurring risk themes or concentrations of risk on a project, it would be helpful if there was a simple way of describing the structure of project risk exposure. A number of existing techniques can be adapted to address this issue, and will allow root causes of risk on a project to be explored.

One approach is to use the standard problem-solving technique of *Root Cause Analysis* (RCA) to identify key risk drivers (defined as events, situations, or conditions that give rise to risk, though they are not themselves risks). The basic RCA process is used routinely in safety and operations analysis to address situations that require resolution, and involves asking the question "Why?" several times until no additional useful information can be obtained. The answer to the last question is then the root cause. A more detailed development of RCA distinguishes between "causes" and "permitting conditions," allowing detailed preventive responses to be developed to break the chain of events that led to the problem.

A more simple technique related to RCA is known as "*Five Whys*," based on the idea that asking the question "Why?" five times will get to the root cause of any event or situation. Either RCA or Five Whys can be applied to the Risk Assessment task, by starting from a statement of risk, and seeking to discover the underlying reason that gives rise to the risk, asking "Why might this risk occur, what would have to happen to cause or allow the risk?" This is undertaken for all risks, and the results are grouped to identify common themes or root causes. After these key risk drivers are identified, management attention can be focused on addressing them to ensure that the risks arising are optimized (i.e., minimizing threats and maximizing opportunities). This can be achieved by a modification of the standard Root Cause Analysis technique, known as Root Cause Projection. The difference is that RCA is used after an event has occurred, and seeks to discover the preexisting causes that gave rise to the event. Risks, however, have not yet occurred, so there is no current situation to analyze using traditional RCA. Instead the project team are required to work backward from identified risks to explore what might happen to cause these risks or allow them to occur.

A more complex approach to analyzing root causes of risk involves production of a *Risk Concept Map*, as shown in Figure 2 of Chapter 4, which indicates relationships between risk drivers (causes), risks (risk situations), and effects (impacts). The effort required to generate the Risk Concept Map is likely to be rewarded with an increased understanding of the relationships between causes and the resulting risks, and analysis of the Risk Concept Map can reveal which of the various causes are the most significant drivers of risk, i.e., those root causes that give rise to the most uncertainty and the highest degree of impact on project objectives.

Another technique that can be used to expose root causes is *Affinity Analysis*, grouping identified risks to determine interdependencies and relationships between them. Derived from a quality management technique, the affinity diagram, or KJ method (named after its author, the Japanese anthropologist Jiro Kawakita, 1920–), has become one of the most widely used of the Japanese management and planning tools. The affinity diagram was developed to discover meaningful groups of ideas within a raw list. It is commonly used to refine the raw results of a brainstorm into something that makes sense and can be dealt with more easily. Its use is recommended when facts or thoughts are uncertain and need to be organized, when preexisting ideas or paradigms need to be overcome, when ideas need to be clarified, and when unity within a team needs to be created. An affinity diagram is created by sorting the original raw list, moving ideas from the list into affinity sets, and creating groups of related ideas. The process takes large amounts of data and organizes it into groupings based on the relationships between the items. It is a creative rather than logical process. During affinity analysis, it is important to let the groupings emerge naturally, using the right side of the brain, rather than using predetermined categories. For Risk Assessment, this usually involves a team process where risk categories are generated into which risks are allocated, thus revealing relationships between risks, and root causes giving rise to groups of risks.

Common Sources of Risk—Using the Risk Breakdown Structure

The previous section describes modifications of existing techniques (Root Cause Analysis, Five Whys, Risk Concept Mapping, Affinity Analysis) that can be used to explore root causes of risk. In addition to these, direct categorization of risks is possible, seeking to expose patterns of risk in terms of both cause and effect.

In any situation where a lot of data is produced, structuring is an essential strategy to ensure that the necessary information is generated and understood. The most obvious demonstration of the value of structuring within project management is the Work Breakdown Structure (WBS), which is recognized as a major tool for the project manager, because it provides a means to structure the work to be done to accomplish project objectives. The Project Management Institute defines a WBS as "a deliverable-oriented grouping of project elements that organises and defines the total work scope of the project. Each descending level represents an increasingly detailed definition of the project work." The aim of the WBS is to present project work in hierarchical, manageable, and definable packages to provide a basis for project planning, communication, reporting, and accountability.

In the same way, risk sources can be organized and structured, to provide a standard presentation of project risks that facilitates understanding, communication, and management. In the simplest case one could produce a list of potential sources of risk, providing a set of headings under which risks can be arranged (sometimes called a risk taxonomy). Many risk prompt lists have been generated that list types of risk to be considered, and that can be used for subsequent categorization. These are often long and unstructured, however, not producing useful information for the Risk Assessment. An alternative is to use a shorter set of categories, and some organizations have developed lists of risk sources for this purpose, for example:

A number of related frameworks exist containing common elements, of which the simplest is PEST (Political, Economic, Social, Technological), sometimes expanded to PESTLE (adding

Legal and Environmental), or even PESTLIED (PEST plus
Legal, International, Environmental, and Demographic).

A leading energy company uses the TECOP framework to structure
sources of risk, including Technical, Environmental, Commercial, Operational, and Political.

CERAM (the Centre for Education and Research in Management)
has developed the SPECTRUM framework for risk sources,
covering Sociocultural, Political, Economic, Competitive,
Technology, Regulatory and legal, Uncertainty and risk,
Market.

However, a simple list of risk sources does not match the richness of
the WBS since it only presents a single level of organization. A better
solution to the structuring problem for risk management would be to adopt
the full hierarchical approach used in the WBS, with as many levels as are
required to provide the necessary understanding of risk exposure to allow
effective management. Such a hierarchical structure of risk sources is
known as a Risk Breakdown Structure (RBS). Following the pattern of
the WBS definition above, the RBS is defined as "a source-oriented
grouping of project risks that organizes and defines the total risk exposure
of the project. Each descending level represents an increasingly detailed
definition of sources of risk to the project." The RBS is therefore a
hierarchical structure of potential risk sources. The value of the WBS lies
in its ability to scope and define the work to be done on the project;
similarly the RBS can be an invaluable aid to understanding the risks faced
by the project.

It is theoretically possible to devise a generic RBS that might apply to
any type of project. However, a generic RBS is unlikely to include the full
scope of all possible risks to every project, so organizations wishing to use
the RBS as an aid to risk management tend to develop their own tailored
versions, perhaps with several different RBS structures applying to particular project types. Large projects may justify development of their own
specific RBS.

A typical RBS structure is shown in Figure 5. This example shows
that at the highest level (RBS Level 0), the source of all risks to the project
might be described simply as "Project risk." The next level of detail is the
major sources of risk to the project, shown in the example figure at RBS
Level 1 as risks arising from Technical, Management, Commercial, or
External sources. RBS Level 2 breaks these into further detail, giving
specific headings under each major source. Individual risks form Level 3 of
the RBS, mapped to one of the Level 2 areas.

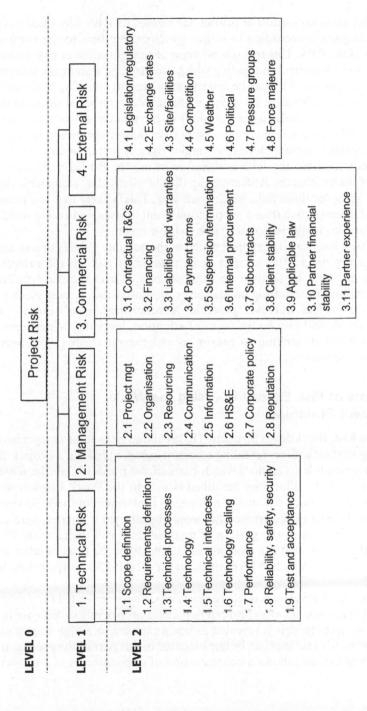

Figure 5 Typical Risk Breakdown Structure (RBS).

LEVEL 0

Project Risk

LEVEL 1

1. Technical Risk

2. Management Risk

3. Commercial Risk

4. External Risk

LEVEL 2

1.1 Scope definition
1.2 Requirements definition
1.3 Technical processes
1.4 Technology
1.5 Technical interfaces
1.6 Technology scaling
1.7 Performance
1.8 Reliability, safety, security
1.9 Test and acceptance

2.1 Project mgt
2.2 Organisation
2.3 Resourcing
2.4 Communication
2.5 Information
2.6 HS&E
2.7 Corporate policy
2.8 Reputation

3.1 Contractual T&Cs
3.2 Financing
3.3 Liabilities and warranties
3.4 Payment terms
3.5 Suspension/termination
3.6 Internal procurement
3.7 Subcontracts
3.8 Client stability
3.9 Applicable law
3.10 Partner financial stability
3.11 Partner experience

4.1 Legislation/regulatory
4.2 Exchange rates
4.3 Site/facilities
4.4 Competition
4.5 Weather
4.6 Political
4.7 Pressure groups
4.8 Force majeure

Once an organization or project has defined its RBS, identified risks can be categorized according to source by allocating them to the various elements of the RBS. This then allows areas of concentration of risk within the RBS to be identified, indicating which are the most significant sources of risk to the project. This can be determined by simply counting how many risks are in each RBS area. However, a simple total number of risks can be misleading, since it fails to take account of the relative severity of risks. Thus one RBS area might contain many risks that are of minor severity, whereas another might include fewer major risks. A better measure of risk concentration within the RBS is therefore to use the P-I Score. Concentration of risks within the RBS areas can then be assessed by comparing the total P-I Score for those risks within each area. This is likely to give a more meaningful perspective than a simple total count of risks, indicating which RBS areas are giving rise to more risk to the project.

After it has been determined which of the possible sources of risk are in fact giving rise to the greatest concentration of risk on the project, attention can then be focused on these sources. Where these are threat sources, this involves putting mitigation or preventive measures in place to protect the project wherever possible. A concentration of sources of opportunity should also be the focus of attention, investing management time and effort in seeking to maximize and exploit them to optimize benefits to the project.

Hot Spots of Risk Exposure—Using the Work Breakdown Structure

While the Risk Breakdown Structure (RBS) allows risks to be categorized according to their source, revealing common causes of risk on a project, it is similarly possible to explore which parts of the project might be most affected by risk, by allocating identified risks into the Work Breakdown Structure (WBS). Each risk can be mapped to the element of the WBS that represents the part of the project that would be affected if the risk were to occur. Examination of the pattern of mapped risks will then expose the parts of the project that are most at risk, using either the total number of risks associated with each WBS element, or the total P-I Score against each element.

After the WBS elements where most risks are concentrated have been identified, management attention can be focused proactively in these areas. Where a project element is revealed as being exposed to a high degree of threats, protective actions can be implemented in this part of the project. If the mapping exercise shows a concentration of opportunities in a project

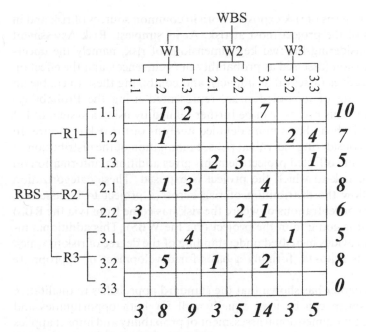

Figure 6 Correlating RBS with WBS.

element, this might be the subject of a cost-benefit study to determine the value of enhancing the project in this area.

A further refinement of this approach is to create a two-dimensional map with the RBS forming one axis and the WBS on the other, as in Figure 6. The distribution of risks can then be plotted to summarize the distribution of both sources (in the RBS dimension) and effects (against WBS). Rows in this matrix with the greatest numbers of risks or the highest total P-I Score indicate the most significant sources of risk (for example, RBS element R1.1 in Fig. 6), and the columns with the highest concentration of risk indicate the parts of the project most at risk (WBS element 3.1 in Fig. 6). Those cells containing most risk also clearly deserve focused management attention.

UNDERSTANDING LEADS TO ACTION

The aim of the qualitative Risk Assessment phase is to prioritize individual risks so attention can be focused on the most significant ones, as well as to determine patterns of risk exposure both in common sources of risk and in those parts of the project most at risk. At its simplest, Risk Assessment involves considering the two key dimensions of risk, namely the uncertainty dimension (assessed as probability of occurrence), and the effect on project objectives (assessed as impact), and combining these to enable an overall ranking of identified risks, represented using the Probability-Impact Grid. This process can be further refined by using a system of P-I Scores, which then allows more detailed analysis of the risk exposure. In addition to considering individual risks, examination of the distribution of risk sources and affected project elements gives additional information on the types of risk to which the project is exposed. These categorization activities allow the distribution of risk across a project to be assessed, looking for concentrations of where the risk has come from (via the RBS) and where it is going to in the project (via the WBS). This additional information provides increased understanding of the degree of risk to which the project is exposed, forming a basis for development of appropriate responses.

This chapter has shown that the standard approaches to qualitative Risk Assessment can be used equally well for both opportunities and threats. The two-dimensional assessment of probability and impact applies in exactly the same way for both types of risk, with the only difference being the definitions of impact: negative for threats and positive for opportunities. Representing these two dimensions on the P-I Grid can also be done in the same way for both threats and opportunities, allowing them to be prioritized for further action, though it is useful to distinguish the two types either through coding or by using a double P-I Grid, with the mirror format concentrating attention on the worst threats and best opportunities in the central zone. The use of a P-I Scoring system can be applied to both opportunities and threats to refine the prioritization process. Similarly, various approaches to categorization work for both upside and downside risk, providing information on the concentration of threats and opportunities. Where categorization indicates common sources of either threat or opportunity, preventive action can be taken. And where areas of the project are revealed as being at risk from significant threat or where there is significant opportunity, proactive action can be taken.

All of these Risk Assessment techniques provide valuable information on the extent of risk faced by the project, resulting in improved understanding of the risk challenge to be managed. The purpose of undertaking Risk Assessment is not, however, simply to create understanding. This is merely a means to a further end, which is to enable development of effective and appropriate responses to risk. If the risk process stops with Risk Assessment, then nothing has changed and the project is as exposed to risk as it was before any risk activities were undertaken. Understanding must lead to action, otherwise Risk Assessment will have been a waste of time. The use of Risk Assessment information as a basis for Risk Response Planning is covered in detail in Chapter 7. This chapter has concentrated on qualitative Risk Assessment as an essential step in the risk process. There is, however, a role on some projects for additional quantitative Risk Analysis before proceeding to response development, and this is the subject of the next chapter.

REFERENCES

Andersen, B., Fagerhaug, T., Anderson, B., Andersen, B. (1999). *Root cause analysis: Simplified Tools and Techniques*. American Society for Quality.

Baccarini, D., Archer, R. (2001). The risk ranking of projects: a methodology. *Int. J. Project Management* 19(3):139–145.

Bartlett, J. (2002). *Managing Risk for Projects and Programmes: A Risk Handbook*. Hook, Hampshire, UK: Project Manager Today Publications.

Bartlett, J. (2002). Using Risk Concept Maps in a project or programme. In: *Proceedings of the 5th European Project Management Conference* (PMI Europe 2002), presented in Cannes, France, 19–20 June 2002.

Boehm, B. W. (1989). *Software Risk Management*. Los Alamitos, CA: IEEE Computer Society Press.

Carr, M. J., Konda, S. L., Monarch, I., Ulrich, F. C., Walker, C. F. (1993). *Taxonomy-Based Risk Identification*. Technical Report CMU/SEI-93-TR-6, ESC-TR-93-183. Pittsburgh, PA: Software Engineering Institute, Carnegie Mellon University.

Chadbourne, B. C. (2001). Root Cause Projection technique for risk management. In: *Proceedings of the 32nd Annual Project Management Institute Seminars*, Symposium (PMI 2001), presented in Nashville, USA, 5–7 November 2001.

Graves, R. (2000). Qualitative risk assessment. *PM Network* 14(10):61–66.

Greenwood, W. (CERAM) (2003). Objective setting: the first critical step in any business change programme. In: *Proceedings of the Business Leadership Conference 2003* (BLC 2003 Europe), held in London, UK, 26 March 2003.

Grimaldi, S., Rafele, C. (2002). Analisi dei rischi nelle attività di progetto. *Impiantistica Italiana* 15(6):69–77.

Grimaldi, S., Rafele, C. (2003). La gestione del rischio: elemento integrante della gestione del progetto. *Logistica Management* 136 (in press).

Hamm, R. M. (1991). Selection of verbal probabilities: a selection for some problems of verbal probability expression. *Organisational Behav. Hum. Decision Processes* 48:193–223.

Hatto, J. (December 2002). Root Cause Analysis and knowledge management: understanding the links. *InfoRM J. UK Inst. Risk Management*: 14–15.

Hillson D. A. (2002). Extending the risk process to manage opportunities. *Int. J. Project Management* 20(3):235-240. [Also in *Proceedings of the 4th European Project Management Conference* (PMI Europe 2001), presented in London, UK, 6–7 June 2001.]

Hillson, D. A. (2002). Structuring a breakdown: the risk management RBS. *Project* 15(7):12–14.

Hillson, D. A. (2002). The Risk Breakdown Structure (RBS) as an aid to effective risk management. In: *Proceedings of the 5th European Project Management Conference* (PMI Europe 2002), presented in Cannes, France, 19–20 June 2002.

Hillson, D. A. (2002). Using the Risk Breakdown Structure (RBS) to understand risks. In: *Proceedings of the 33rd Annual Project Management Institute Seminars, Symposium* (PMI 2002), presented in San Antonio, USA, 7–8 October 2002.

Hirsch, K., Wallace, D. (2000). *Step-by-Step Guide to Effective Root Cause Analysis*. Marblehead, MA: HCPro Inc.

Lansdowne, Z. F. (1999). Risk matrix: an approach for prioritising risks and tracking risk mitigation progress. In: *Proceedings of the 30th Annual Project Management Institute Seminars, Symposium*, presented in Philadelphia, USA, 11–13 October 1999.

Latino, R. J., Latino, K. C. (1999). *Root Cause Analysis— Improving Performance for Bottom-Line Results*. Boca Raton, FL: CRC Press.

Lichtenstein, S., Newman, J. R. (1997). Empirical scaling of common verbal phrases associated with numerical probabilities. *Psychonom. Sci.* 9.

Pritchard, C. (2001). How high is 'high'? Setting risk process metrics. *Project Manager Today* 13(10):22–24.

Pritchard, C. (2002). Your finger on the trigger—the thresholds of risk and risk strategies. *Project Manager Today* 14(6):12–14.

Project Management Institute. (2002). *A Guide to the Project Management Body of Knowledge (PMBoK®)* 2000 ed. Philadelphia: Project Management Institute.

Project Management Institute. (2002). *Practice Standard for Work Breakdown Structures*. Philadelphia: Project Management Institute.

Simon, P. W., Hillson, D. A., Newland, K. E., eds. (1997). *Project Risk Analysis, Management (PRAM) Guide*. High Wycombe, Bucks UK: APM Group.

Ward, S. C. (1999). Assessing and managing important risks. *Int. J. Project Management* 17(6):331–336.

Williams, T. M. (1996). The two-dimensionality of project risk. *Int. J. Project Management* 14:185–186.

6
Quantitative Analysis

The qualitative Risk Assessment phase discussed in Chapter 5 takes identified risks and lists them in priority order for further action and may also consider grouping risks into categories to focus management attention on hot spots of exposure. Each risk is treated as distinct, independent, and separate from each other risk. While this is an important step to develop understanding both of individual risks and of the overall risk exposure facing the project, it is a necessary simplification of the situation.

Risks do not occur on projects singly or in related groups, but in random collections. Neither are risks unconnected from each other, since the occurrence of one risk can make another risk more or less likely, and could increase or decrease its impact. The occurrence of a risk can even create new risks or preclude others.

The greatest weakness of the qualitative Risk Assessment phase is that it addresses risks one at a time to prioritize them, and puts them into similar groups to expose patterns of risk. An accurate understanding and analysis of the potential impact of uncertainty on project objectives therefore requires an approach that deals with the combined effects of risks occurring randomly, and includes consideration of interrisk effects.

This is the purpose of the quantitative Risk Analysis phase, which aims to build a model representing the project into which the various identified risks can be incorporated, thus allowing a synthesized view of the overall effect of risk on the project, which should of course include both threats and opportunities. A number of different techniques exist for building such simulation models, the most popular of which are Monte Carlo models, decision trees, and influence diagrams (although other more complex modeling approaches are possible, including system dynamics, neural networks, etc.). This chapter concentrates on Monte Carlo analysis, since it is the most commonly used risk analysis technique, though many of the principles discussed here apply equally to use of other analytical modeling approaches. Many practitioners of quantitative Risk Analysis are taking account of opportunities alongside threats when they construct their risk models, but this is often done subconsciously or intuitively, without a clear recognition of the place that opportunity deserves in this analytical process. After discussing the typical approach to Risk Analysis, this chapter considers how to ensure that opportunities are included in the process explicitly.

INTRODUCING MONTE CARLO

Monte Carlo simulation is popular because it is based on a well-understood set of statistical sampling rules, and does not require specialist understanding. It involves development of a risk model that is usually based on an existing project plan, rather than requiring a bespoke model to be constructed from first principles, thus building on a basis that is understood and accepted by the project stakeholders. The basic Monte Carlo approach is intuitively self-evident, since it involves taking the project plan and adding known uncertainty to analyze resulting deviations from plan.

The underlying key concept of Monte Carlo simulation is the replacement of single-point deterministic values in a project plan with ranges to reflect uncertainty. A typical project plan includes fixed values for each element or activity, describing the duration, cost, resource level, etc., for that activity, and allowing the overall project duration, cost, or resource requirement to be simply determined. Reality is of course different, since the precise values for these are not fixed, and activity duration, cost, and resource level are variables. For example, the actual duration of a given project activity might vary from the value in the original plan for many reasons, including estimating inaccuracy or bias, unplanned work or rework, and, of course, the effect of risks (both positive and negative). Consequently the reflection of activity duration (or cost, or resource, or any other activity variable) by a single number in the project plan is misleading. A Monte Carlo risk model allows such single-point estimates to be replaced by ranges to reflect the associated uncertainty. The simulation is then performed by taking multiple random iterations through the risk model, sampling from input ranges. Each iteration produces one feasible outcome for the project, calculated from a sample of values drawn from the input data. Multiple iterations generate a set of results reflecting the range of possible outcomes for the project, which reflect the best case, the worst case, and all conditions in between. Results are typically presented as an S-curve, a plot of the range of possible outcomes against the cumulative probability of achieving a given value. (S-curves are described later in this chapter.)

WHY QUANTIFY?

Using simulation to reflect and analyze the effect of risk on the overall outcome of a project has a number of benefits, including:

- A means of analyzing the combined effect of risks together on a project, rather than treating them individually
- Describing risks quantitatively, using numbers or ranges for probability and impacts, instead of ambiguous descriptive terms such as High or Low
- Consistency in analysis, since the operation and output of a simulation model is independent of the person running the analysis, and is not subject to subjective preconceptions and bias
- The ability to flex the model to analyze different scenarios and alternatives to the base case, allowing exploration of a range of options for addressing risk
- The power to reflect a degree of complexity that exceeds what can be understood by a single person or held in his memory, allowing development of a sophisticated model of reality that can accurately predict outcomes
- Presentation of project outturn as a range of possible outcomes rather than a single point

There are also, however, shortcomings in using a simulation approach for addressing risk, including:

- Use of software tools which, although they may have good functionality and capability to support detailed analysis of risk, are an additional cost to the project, are likely to require staff training if they are to be used effectively, and require integration with other project tools.

- Requirement to interpret analysis outputs, which may need some understanding of statistical principles to avoid misinterpretation.
- The danger of spurious precision, since computer-based tools naturally produce outputs to many decimal places, suggesting a degree of precision that is unlikely to be justified by the input data.
- A related danger of spurious credibility, giving too much credence to model outputs without applying sufficient critical thought or judgment to the results.
- The use of specialized tools can result in dependence on an "expert" to run the analysis, performing tasks that are not understood by the project team, leading to a degree of separation and loss of ownership.
- Some project stakeholders may feel they are able to abdicate responsibility, believing that reporting a potential risk means they do then not have to take action.

The existence of both strengths and weaknesses in the Risk Analysis approach mean that care should be exercised when implementing this phase of the risk process. In fact, most risk management guidelines recommend that quantitative Risk Analysis should be treated as an optional phase of the process, and not used for all projects. Apart from avoiding the shortcomings listed above, the other main reason for not using quantitative Risk Analysis is that often the qualitative Risk Assessment approach is entirely adequate to allow effective management of identified risk. Simply considering the likelihood and effects of each risk can lead to development of risk responses that deal appropriately with the risk exposure faced by the project, without the need for more complex quantitative Risk Analysis. On many low-risk projects the degree of risk exposure does not warrant use of sophisticated quantitative analysis techniques, since a simple qualitative assessment will provide sufficient understanding of the risks to enable them to be addressed. In other cases the time or resource available for risk management may preclude a more detailed quantitative analysis, and it may only be possible to make a quick qualitative assessment.

It is vital to understand the reason why quantitative Risk Analysis might be undertaken on a particular project. In all cases Risk Analysis is merely a means to an end, providing useful information to enable informed decisions to be made in response to the degree of risk faced by the project. Risk Analysis essentially produces numbers, but the project team need strategies for dealing with risk. Indeed what is required is not Risk Analysis, but Risk Management. Skeptics might say that the purpose of quantitative Risk Analysis is to predict cost and schedule overruns well in advance. Of course, in reality the purpose is (or should be) to increase understanding of the risk exposure to allow effective action to be taken proactively. It is very easy when implementing quantitative Risk Analysis to get seduced into the intricacies of building, running, and interpreting a risk model, and to forget that the purpose of the analysis is to support the decision-making process.

Before making the decision to use Risk Analysis on a particular project, the project team should consider whether it is justified by the level of risk exposure, whether quantitative data are required in addition to qualitative assessment information, and whether the necessary support infrastructure is in place (tools, training, expertise, etc.). Clear objectives should be set for the use of quantitative Risk Analysis, being aware of the potential shortcomings, and the project team should be prepared to make full use of the analysis results in the project decision-making process. If these factors are not considered, there is a danger that Risk Analysis will be undertaken for the wrong reasons and that it will fail to deliver the expected benefits. The selected approach to Risk Analysis should be recorded in the Risk Management Plan (see Chapter 3), together with reasons to support the decision on whether or not to include this phase in the risk process.

REALISTIC INPUT DATA FOR ANALYSIS

Whatever analysis approach is used, a prerequisite is to produce data in a quantitative format as input to the risk model. The "garbage in/garbage out" principle applies here, since the quality and usefulness of any analysis outputs depend on the quality of input data. Particularly when generating hard numbers as inputs to Risk Analysis, it is vital to know where those numbers originate. Simply "guesstimating" is dangerous, as invalid input data will result in misleading output results.

The recommended approach to risk management includes qualitative Risk Assessment as a mandatory phase, since it is clearly important for every project team to understand the risks to which the project is exposed. Given the existence of a number of identified risks that have been assessed in this preceding phase, it makes sense to use this information as the basis for developing input data for a risk model.

The usual approach to incorporate uncertainty into a risk model involves mapping identified risks into the project plan, determining which elements of the plan are affected by which risks (see Fig. 1). When using Monte Carlo simulation, this involves the following steps :

1. *Generate a risk model based on the current project plan.* For a Risk Analysis examining schedule or resource risk exposure, the project critical path network should be used as a basis for the risk model. Cost Risk Analysis is usually based on the project cost breakdown structure, often held in a spreadsheet, although this too can be based on the critical path network where activities are fully costed. An integrated Risk Analysis can also be performed, addressing both schedule, resource, and cost within the same risk model. Although the risk model is derived from the project plan, it is not usually identical with it. Modifications include changing the level of detail within the project plan to enable risks to be modeled appropriately. Where the project plan includes routine elements with little or no associated risk (for example, scheduled management review meetings or the regular reporting cycle), the level of detail in the risk model can be reduced. However, detail should be retained or even increased in the high-risk areas of the project plan where many risks have been identified, enabling their effect to be modeled. The level of risk associated with different areas of the project plan can be determined from risk categorization undertaken during the Risk Assessment phase (see Chapter 5).

2. *Reflect alternative logic paths using stochastic branches.* There may be some areas of the project where the precise detail of the work is not yet fully defined or scoped, or where different options remain. In the normal project plan it is necessary to assume which approach will be taken, and perhaps to run alternative versions of the plan to examine the options. A risk model can, however, include alternative logic paths to reflect the uncertainty relating to future decisions. In the same way that the project team might decide which option to take when they have made some progress on the project, the simulator can be given the freedom to select between alternative logic paths on different iterations through the model. This is done using stochastic branches, either probabilistic or conditional (see examples in Fig. 2), to model optional activities that may form part of the project under certain circumstances.

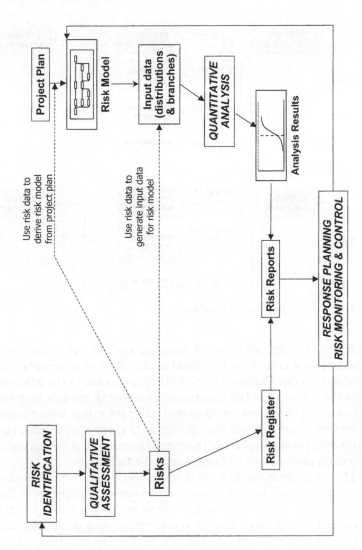

Figure 1   Mapping risks during Risk Analysis.

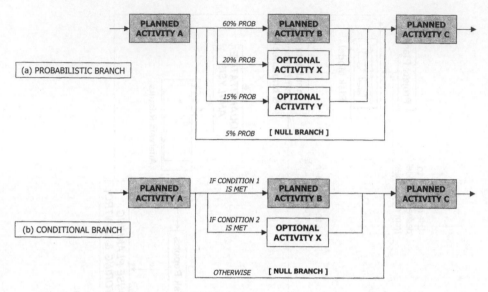

Figure 2 Examples of stochastic branches.

3. *Identify targets within the risk model against which the uncertainty can be measured.* Risk is defined as the effect of uncertainty on objectives (see Chapter 1), and both elements need to be reflected in the risk model. The uncertainty is modeled through mapped risks, and objectives are identified as target values whose feasibility is tested through the simulation. Targets may include key time milestones, especially the overall project duration and completion date, as well as budget and resource targets.

4. *Map risks from the Risk Assessment phase into the risk model.* Some risks will be easily mapped to individual elements in the risk model, particularly where the scope of the risk is clearly defined. Other risks may be more generic and could exert an influence on many model elements. Some risks identified during the Risk Assessment phase may be excluded from a particular analysis (e.g., risks with pure performance impact may not be included in a cost risk analysis), and these exclusions should be recorded for future examination, together with all other assumptions made during production of the model.

5. *Identify effect of mapped risks on values of affected model elements.* This involves replacing single-point estimates in the

project plan with ranges in the risk model, typically producing a three-point estimate to reflect the best case, most-likely value, and worst case. Other distributions can be used where it is not appropriate or possible to generate a three-point estimate, including a uniform distribution (only defining best and worst cases), discrete distributions (noncontinuous specific values), and various curves (normal, beta, gamma, lognormal, exponential, etc.). The range of values used for elements in the risk model should include taking account of planned responses where these are known. When Risk Analysis is being undertaken prior to the Response Planning phase, the initial risk model obviously cannot include responses, but later updates of the analysis should incorporate the expected effect of risk responses into the model.

6. *Introduce specific model constructs for key risks.* It may not be appropriate to represent some risks by mapping them to existing elements within the risk model, particularly where such risks describe unusual events outside the planned activities. These risks may be included in the risk model using stochastic branches, to reflect the fact that a particular risk might occur driven either by its probability of occurrence or as a result of other uncertain conditions being met. In the example of a probabilistic branch shown in Figure 2a, the impact of a risk could be included as the "optional activity," with its probability of occurrence controlling the frequency with which the branch is sampled during the simulation. A risk can also be modeled as a conditional branch [Fig. 2(b)] where the impact is determined by fulfillment of a defined condition.

7. *Add dependency/correlation groups to reflect risk relationships.* Although it may sometimes seem like it to the project team, it is not true that things happen completely by chance on a project. There are limits to randomness, particularly later in a project, when the degrees of freedom are constrained by what has happened before. This needs to be reflected in the risk model, otherwise the Monte Carlo simulation will assume complete randomness for all uncertain variables throughout the risk model. Absence of correlation will reduce the spread of results calculated during the simulation, owing to the operation of the central limit theorem, with random uncertainty in uncorrelated elements canceling out. In reality, risks are interdependent: if one risk occurs it can make another risk more or less likely,

with a greater or lesser impact, or even preclude one or more risks from happening while introducing new risks. Planned activities are also related, with earlier performance setting trends that persist later. These relationships need to be included in the risk model, creating dependency through correlation groups to link activities and risks that can affect each other. Such links may be driven by the existence of common causes or external dependencies, or where a single risk affects several elements of the model, or where activities are performed by the same resource. In these cases the ability of the simulator to sample randomly needs to be constrained. A correlation group identifies elements in the model where sampled values are related, either positively or negatively, and uses a correlation coefficient (between −1 and +1, or from −100% to +100%) to model the strength of the relationship.

All of these steps are important to ensure that the risk model realistically reflects the effect of uncertainty on the project. The key to producing an accurate model, however, is step 5, generating input data to reflect the effect of mapped risks on elements in the risk model. This needs to be done using a structured approach, taking information about mapped risks from the Risk Assessment phase.

OPTIMISTIC OR OPPORTUNISTIC?

The usual approach to generating input data for a Monte Carlo risk model starts with the estimates contained in the existing project plan, and modifies these to reflect the effect of mapped risks. The following discussion focuses on production of three-point estimates since these are most commonly used, but the principles apply equally to other types of distribution.

The typical process for generating a three-point estimate for a risk model starts with the value from the project plan. This value often includes some form of contingency to take account of unknown or unidentified risks. The first step in the three-point estimating process is to remove that contingency to leave the raw estimate. That estimate is then varied to take account of the effect of risks that might affect the element under consideration, to produce the best case and worst case. The best case is usually taken to be the raw estimate without contingency, and in the worst case the full effect of all mapped risks is assumed to occur, possibly with addition of "real contingency" to account for the effect of unidentified risks. The most-likely value is then determined within the range of best-worst, by considering the probabilities of mapped risks and deciding which are likely to occur and at what level of impact.

This "adjust-from-plan" method is commonly used to produce three-point estimates, but suffers from a major disadvantage where the risk process focuses only on identification and management of threats. This is illustrated in Figure 3, which shows the effect of a number of threats on a project plan, whose impact would cause a shortfall in project performance if the threats occurred. Responses will of course be designed to tackle those threats, but it is most unlikely that identified responses will be fully effective, with the end result that only a partial recovery will be achieved. This creates a one-way street for the project, with threats taking it away from achieving the plan, and the best possible case being simply to get back to the plan. Using the language of the "Roundabouts and Swings" poem ("Losses on the roundabouts means profits on the swings," P. R. Chalmers, 1925), this approach sees only roundabouts.

With only threats included in the risk process, the best-case value in a three-point estimate usually equals the planned value with no contingency. This, however, ignores the existence of opportunities to improve on project performance, to create increased benefits, to raise productivity levels, avoid rework, use new approaches to reduce time and cost, etc. Without opportunities the best case is bound to be an underestimate, representing only an optimistic view of the "least bad" situation.

Inclusion of opportunities in the risk process, however, leads to a different approach for generating three-point estimates, as reflected in

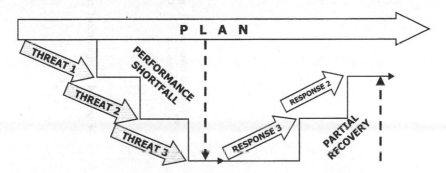

Figure 3 Effect of threats on the plan.

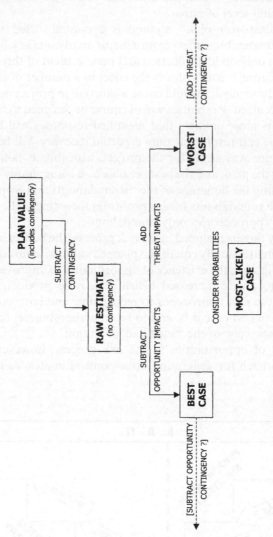

Figure 4   Adjustment estimating method including opportunities.

Figure 4. Here the best case is not only the raw estimate of the planned value excluding contingency. Instead the best case also includes the effect of any identified opportunities that might affect the element under consideration, allowing for the possibility to improve on the raw estimate. This produces a wider spread of values for the three-point estimate, which more realistically reflects the true range of possible uncertainties associated with a particular element in the risk model. Figure 5 illustrates this, with the central range of the distribution representing the area of estimating uncertainty (i.e., the risk that the original planned value may be wrong by some small degree, plus or minus), whereas the extremes of the distribution represent opportunities to significantly improve on the planned value (left-hand side) or threats that the plan will be significantly exceeded (right-hand side).

Returning to the imagery of "Roundabouts and Swings," including opportunities in the three-point estimating process introduces the possibility of some swings to compensate for the undoubted existence of threat roundabouts. Taken together with threats, opportunities negate the one-way street that leaves a project facing failure with the only option being how bad that failure might be. With both threats and opportunities the project has a better chance of remaining on target and meeting project objectives.

The discussion above has focused on use of opportunity and threat data when producing three-point estimates. It is, however, possible to use a similar approach for other distribution types (uniform, discrete, or curves),

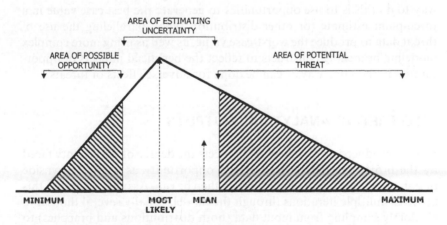

Figure 5 Regions of a three-point estimate.

ensuring that any possible upside impact is explicitly reflected in the optimistic side of the distribution.

In addition to using opportunity data when producing estimates for input ranges, there are other ways of including opportunities within the risk modeling process using the stochastic branches mentioned in step 6 above, including both probabilistic and conditional branches (see Fig. 2). A probabilistic branch presents two or more alternative logic paths within the risk model, allowing the simulator to select different paths for different iterations through the model. Each path in the branch has an associated probability of occurring, and the frequency of selection during multiple iterations is driven by this probability. For a conditional branch, the path followed by the simulator for any particular iteration is determined by whether or not a given condition is satisfied. An opportunity can be reflected as one of the options in a stochastic branch in one of three ways:

> Using a null or empty branch to model the opportunity to remove elements of planned work from the project plan
> Using a shorter or cheaper branch to model alternative options with reduced duration or cost compared to the basic project plan
> Using a negative lag on a branch to model the opportunity to start a succeeding activity early (note that not all software tools allow this type of construct)

It is clearly important to include information on both threats and opportunities when developing input data for a risk model, to avoid a one-sided perspective on the effects of the risks facing the project. The simplest way to do this is to use opportunities to generate the best case value in a three-point estimate (or other distribution type), paralleling the use of threat data to produce the worst-case value, as well as using more complex modeling branching constructs to reflect the beneficial impacts of opportunities in the same way as can be done for adverse effects of threats.

INTERPRETING ANALYTICAL OUTPUTS

Having produced a risk model that reflects the degree of uncertainty faced by the project, including both upside opportunity as well as downside threat, the Monte Carlo simulation can then be run. As outlined above, this involves multiple iterations through the model, usually several thousand, randomly sampling from input data (both distributions and branches) to determine the range of possible outturns for the project. Outputs are usually generated for the overall project, describing the range of project durations or completion dates that might arise given the occurrence or not of identified threats and opportunities, or the range of total project cost or resource requirement. It is also possible to produce outputs for a subset of the project, for example analyzing the uncertainty associated with a particular milestone or cost center or resource type.

S-Curves

The main output from a Monte Carlo simulation is a cumulative probabilistic distribution function known as the S-curve. This may be supported by a histogram that presents the incidence with which each particular result was obtained. An example is given in Figure 6, showing a range of calculated end dates for a project. In this figure, the histogram incidence plot (gray bars) represents the number of times any given date was calculated as a result of a simulation iteration, measured against the left-hand axis. The solid line is a cumulative plot of these data, giving the percentage likelihood of completing the project on or before a given date, plotted against the right-hand axis (0–100%).

The example in Figure 6 shows 100% certainty of completing the project on or before 6 August, since all the calculated results are earlier than this maximum date. There is no chance (0%) of completing earlier

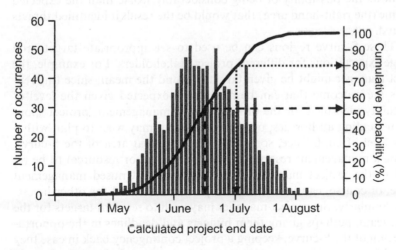

Figure 6 Example S-curve.

than the calculated minimum of 22 April. The mean value (shown in Fig. 6 as a dashed line) represents the completion date that would be expected given the input data with the level of risk included in the model, namely 15 June. The mean is not usually the same as the 50% point (which would be the statistical median value), but is the average of all the calculated values. When results are asymmetrical the mean will be different from the median, and in a right-skewed S-curve like that in the figure, the percentage associated with the mean will be > 50%.

The percentage values read from an S-curve can also be described as "confidence levels," since they represent the chance of meeting a particular value. Thus in Figure 6, if the target date were 1 July the project team could have 80% confidence of completing their project on time, as shown by the dotted line; i.e., there is only a 20% chance of the project finishing later than this date.

In the same way that regions can be identified on an input data three-point estimate that represents normal "estimating uncertainty" and areas of significant opportunity or threat (Fig. 5), similar regions can be identified on the S-curve that represents possible project outcomes, as shown in Figure 7. Here some variation around the mean might be expected in the normal course of events as a result of the usual level of uncertainty associated with undertaking project activities. But the S-curve indicates that the project could perform significantly better than this (the left-hand region of the curve) if identified opportunities are exploited. The S-curve also shows the possibility of being considerably worse than the expected outcome (the right-hand area) that would be the result if identified threats occurred.

These S-curve regions can be used to set appropriate targets and manage expectations for different project stakeholders. For example, the project manager might be given targets around the mean, since this represents the outcome that can be reasonably expected given the level of uncertainty identified for the project. Senior management, project sponsors, or funding authorities, on the other hand, may wish to plan with a higher level of confidence, somewhere in the threat area of the S-curve, creating a management reserve (of finance or time or resource) to be released to the project manager if the need arises. Unused management reserve, of course, can be returned to profit or released for other investments. Similarly, the project manager may wish to set tight targets for the project team, perhaps giving them budgets and deadlines in the opportunity region of the S-curve, keeping a project contingency back in case they are unable to meet these targets.

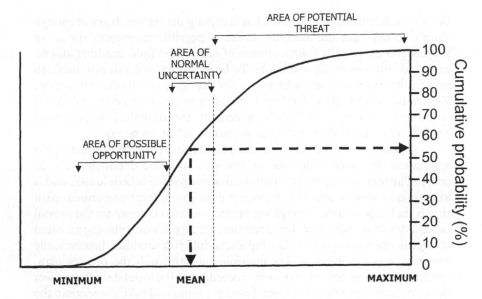

Figure 7 Regions of an S-curve.

So, for example, if the project manager is given a target budget at the mean value of say 55%, senior management might wish to commit funds against the project at the 85% confidence level to cover the threat of possible overruns. The project manager may similarly give project team members targets in the opportunity region, with say only 35% chance of being met, recognizing that this is a challenging stretch target that may not be achieved. For this project, the project contingency held by the project manager would then equal the difference between the 35th and 55th percentiles on the S-curve, and the management reserve held outside the project would be set at the difference between 55% and 85%.

Criticality and Cruciality/Sensitivity

The S-curve allows the combined effect of both estimating uncertainty and explicit risks on project outturn to be analyzed. The likelihood with which project objectives might be achieved can be expressed as a percentage confidence level, and the range of possible outcomes can be explored. Having created a view of the overall project, however, it is useful to be able to look behind the headlines to examine what is driving the results. It is not enough simply to say that the analysis reveals a possible six-month spread in project end dates with a 40% chance of hitting the final deadline and an expected outcome of six weeks late. To be useful the analysis also needs to give information on the underlying risks that give rise to this prediction. Remembering that Risk Analysis is merely a means to an end, and that it needs to lead to effective Risk Management, the analytical outputs must provide a diagnosis that enables appropriate action to be taken.

A number of other outputs can be obtained from the Monte Carlo simulation that give additional information in more detail than the S-curve. The first only relates to quantitative analysis of schedule risk, and is known as *criticality analysis*. A project plan has at least one critical path that is the longest route through the project and that determines the overall project duration, with any change on the critical path introducing an equal change in the whole project. During a schedule risk analysis, however, the Monte Carlo simulator makes multiple runs through the project plan, randomly varying activity durations according to the input data that reflect the uncertainty and mapped risks. Some activities will take longer than the planned duration while others will be shorter. As a result the project critical path is likely to vary during the simulation as previously critical activities might be completed in a shorter time while other noncritical activities are extended. In fact, during the many iterations of a risk model a number of alternative critical paths might be followed.

It is possible to calculate a *criticality index* for each activity in the risk model, defined as the number of times that activity appears on the critical path, usually expressed as a percentage of the total number of iterations. Thus an activity that is always critical has a criticality index of 100%, while one that can never be on the critical path has zero criticality. The activities of interest are those with criticality between 1 and 99%, which might become critical under certain circumstances. Ranking activities by criticality index highlights those activities most likely to drive the overall project duration and completion date, and which therefore require focused risk management attention. By concentrating on the threats and opportunities mapped against high criticality activities, the degree of schedule risk can be reduced effectively.

A second useful analysis of detailed risk drivers relates the degree of variability in a particular element of the risk model with the variation in the overall project outcome. It can be applied to any type of risk analysis, including schedule, cost, resource, etc. This factor was first described in 1992 by Williams and Bowers, and termed *"cruciality,"* expressed as a correlation coefficient (between −1 and +1) indicating the relationship between each activity or risk and the total project. (Note that in recent years this same correlation factor has come to be known as "sensitivity" among many risk practitioners, and it is also called this in some risk analysis tools that calculate it. Some people who use "sensitivity" in this way define another different factor that they somewhat confusingly also call "cruciality," calculated by multiplying sensitivity by criticality, though of course this only applies to schedule risk analysis. When used in this latter sense, a high cruciality factor indicates activities or risks whose variation has the biggest effect in driving the overall project end date and duration.)

Elements with high correlation in the cruciality/sensitivity factor are key drivers of risk, since a large change in the element produces a correspondingly large change in the overall project. This is equally true of both threats and opportunities, since highly crucial/sensitive threats have a large adverse effect on the overall project, while highly crucial/ sensitive opportunities have the biggest overall upside impact.

As for criticality analysis, elements in the risk model can be ranked by cruciality/sensitivity to indicate which are the most significant causes of risk to the overall project. This information is often presented graphically as a so-called Tornado Chart (see Fig. 8 for an example) to highlight the major risk drivers. Activities and risks with high cruciality/sensitivity

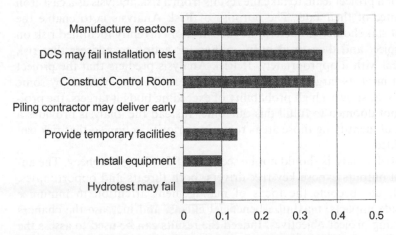

Figure 8   Example Tornado Chart showing cruciality/sensitivity.

should be treated with priority when determining areas for further risk management attention and action.

ANSWERING THE "SO WHAT?" QUESTIONS

Risk Analysis using Monte Carlo simulation provides a powerful method to analyze the combined effect of risks on project objectives, including both threats and opportunities. The input data should be derived from risks that have been identified and assessed in previous phases of the risk process, and that are therefore understood. Output from a risk model allows the range of possible outcomes to be predicted, from the best case where no threats occur but opportunities do happen, through to the worst case where there are only problems arising from unmanaged threats without the beneficial effects of exploited opportunities. It also indicates the best opportunities and the worst threats, with the biggest potential effects on the overall project.

Despite the power of Risk Analysis techniques in predicting possible future outcomes for projects based on identified risk, the outputs obtained from an analysis are not an infallible statement of the inevitable. Simulation results indicate the implications for the project of the input data provided for the model, allowing the project stakeholders to look into the future and foresee the likely consequences of their current actions from the perspective of a particular point in time. It would, however, be most unwise for a project team to take the results from a risk analysis as a cast-iron guarantee of their fate. The purpose of Risk Analysis is to enable the project stakeholders to determine the effect of currently identified risk on the project and then to identify options for action to address that risk and deal with it appropriately. If Risk Analysis predicts that the project cannot meet its target completion date and is likely to overrun by some months, or shows a high probability of exceeding budget targets, the project is not doomed to fulfill this outcome. Instead the analysis provides a means of identifying those areas requiring priority action so that risk can be tackled.

Risk Analysis should not be seen as self-fulfilling prophecy. The analytical outputs expose key risk drivers, both threats and opportunities, that should become the focus of management attention to minimize downside impacts, maximize beneficial effects, and increase the chances of meeting project objectives. Indeed the results can be used to assess the effect of resolving particular risks, by successively removing them from

the risk model, as shown in Figure 9, where S-curves are plotted in a so-
called "onion-ring diagram" to show the influence of particular individual risks on overall project outturn. A similar approach can be used
to test the effectiveness of proposed responses, by comparing preresponse
and postresponse S-curves and other analytical outputs. Analytical results
from the model can be used to answer a range of "So what?" questions,
including:

> What could the risks we have already identified and assessed do to the
>   project as a whole?
> What is the best that we can expect from this project, if we capture all
>   the opportunities and avoid all the threats?
> And how bad might it be if all the threats turn into problems and we
>   miss all the opportunities?
> Within that range of uncertainty, and given our current level of risk,
>   where should we expect the project to end up?

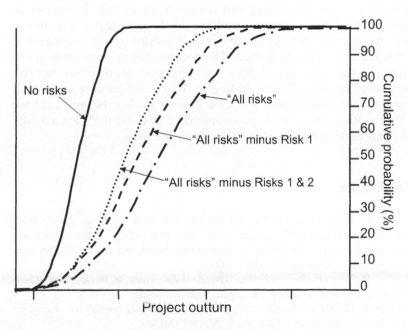

Figure 9 Onion-ring diagram of successive S-curves.

How likely are we to meet project targets, such as duration and
completion date, interim milestones, budget at completion,
profitability, return on investment, etc.?
Which of the risks are the main drivers of overall project uncertainty?
The best opportunities? The worst threats? Where should we
concentrate our risk management efforts?
What could we realistically hope to achieve by managing risks on this
project? What are our options for action? How much difference
can we make?

Risk Analysis certainly provides a powerful method to analyze the
effect of risk on a project, and it needs to explicitly include both threats and
opportunities if it is to accurately reflect the situation faced by the project.
But it is important to beware of overreliance on the outputs from a risk
model. Instead the project stakeholders need to be prepared to challenge
the assumptions underlying the model and its input data, understand the
significant drivers that have the greatest influence on the outcome, and
validate the results to be sure that they make sense.

Although Risk Analysis offers much to support the project stakeholders in their understanding and management of risk, it remains an
optional part of the risk process, and is unlikely to be required or appropriate on every project. Even where it is used, it must be seen as a means to
an end, and not as an end in itself. The point of the risk process is not to
analyze risk but to manage it. Risk Analysis produces numbers, but risk
management requires effective strategies. This is the purpose of the next
phase of the risk process, Risk Response Planning, when the results of both
qualitative Risk Assessment and quantitative Risk Analysis are actually
used to make decisions on what to do about the risks.

REFERENCES

Davis, J. P., Hall, J. W. (1998). Assembling uncertain evidence for decision
making. In: Babovic, V., Larsen, L. C., eds. *Proceedings of the Third International Conference on Hydroinformatics (Hydroinformatics 98)*. Rotterdam,
Netherlands: Balkema, pp. 1089–1094.

Eastaway, R., Wyndham, J. (1998). *Why do buses come in threes? The hidden
mathematics of everyday life*. London: Robson Books.

Garvey, P. R. (2000). *Probability methods for cost uncertainty analysis: A systems
Engineering Perspective*. New York: Marcel Dekker.

Gill, H., Hillson, D. A. (1998). The role of quantitative risk analysis. *Project*
11(3):26–27.

Graves, R. (2001). Open and closed: the Monte Carlo model. *PM Network* 15(12):48–52.

Gray, N. S. (2001). Secrets to creating the elusive 'accurate estimate'. *PM Network* 15(8):54–57.

Grey, S. (1995). *Practical Risk Assessment for Project Management.* New York: John Wiley.

Haimes, Y. Y. (1998). *Risk Modelling, Assessment and Management.* New York: John Wiley.

Hall, J. W., Blockley, D. I., Davis, J. P. (1998). Non-additive probabilities for representing uncertain knowledge. In: Babovic, V., Larsen, L. C., eds. *Proceedings of the Third International Conference on Hydroinformatics (Hydroinformatics 98).* Rotterdam, Netherlands: Balkema, pp. 1101–1108.

Hopkinson, M. (2000). Project plans and planning—the ideal versus the reality. *Project* 13(6):28–29.

Hopkinson, M. (2001). Schedule risk analysis: critical issues for planners and managers. In: *Proceedings of the 4th European Project Management Conference,* (PMI Europe 2001), presented in London, UK, 6–7 June 2001.

Hulett, D. T. (1999). Schedule risk analysis simplified. In: *Proceedings of the 30th Annual Project Management Institute Seminars, Symposium,* presented in Philadelphia, USA, 11–13 October 1999.

Hulett, D. T. (2000). Project scheduling risk analysis: Monte Carlo simulation or PERT? *PM Network* 14(2):43–47.

Hulett, D. T. (2001). Integrating analysis of both schedule and cost risk. In: *Proceedings of the 4th European Project Management Conference* (PMI Europe 2001), presented in London, UK, 6–7 June 2001.

Hulett, D. T. (2002). Integrated cost/schedule risk analysis. In: *Proceedings of the 5th European Project Management Conference* (PMI Europe 2002), presented in Cannes, France, 19–20 June 2002.

Hull, J. K. (1990). Application of risk analysis techniques in proposal assessment. *Int. J. Project Management* 8(3):152–157.

Kuchta, D. (2001). Use of fuzzy No.s in project risk (critically) assessment. *Int. J. Project Management* 19(5):305–310.

Lewis, A. (2000). Analysis: ten tips to make it work. *Project Manager Today* 12(7): 16–18.

Mars, L. A. (2000). Contingency, risk and ensuring quality. *PM Network* 14(12): 56–59.

Pascale, S., Troilo, L., Loranz, C. (1998). Risk analysis: how good are your decisions? *PM Network* 12(2):25–28.

Roberts, B. B. (2001). The benefits of integrated quantitative risk management. In: *Proceedings of the 12th Annual International Symposium of the International Council on Systems Engineering* (INCOSE), held in Melbourne, Victoria, Australia on 1–5 July 2001.

Ruskin, A. M. (2000). Using *unders* to offset *overs. PM Network* 14(2):31–37.

Schuyler, J. (2000). Capturing judgements about risk and uncertainties. *PM Network* 14(7):43–47.

Schuyler, J. (2000). Project risk management by the numbers. *PM Network* 14(9): 69–73.

Schuyler, J. (2001). *Risk and Decision Analysis in Projects*. 2d ed. Philadelphia: Project Management Institute.

Sweeting, J. (2000). Risk analysis without Monte Carlo simulation. *Cost Engineer* 38(3):17–20.

Vose, D. (2000). *Risk Analysis: A Quantitative Guide*. 2d ed. Chichester, UK: John Wiley.

Ward, S. C., Chapman, C. B. (1991). Extending the use of risk analysis in project management. *Int. J. Project Management* 9(2):117–123.

Weiler, C. (1998). Risk-based scheduling and analysis. *PM Network* 12(2):29–33.

West, K. (2001). Estimating to minimise loss. *PM Network* 15(4):42–45.

Williams, T. M. (1992). Critically in stochastic networks. *J. Operational Res. Soc.* 43(4):353–357.

Williams, T. M. (1992). Practical use of distributions in network analysis. *J. Operational Res. Soc.* 43(3):265–270.

Williams, T. M. (1993). What is critical? *Int. J. Project Management* 11:197–200.

Williams, T. M. (2002). *Modelling Complex Projects*. Chichester, UK: John Wiley.

7
Planning Responses

Previous steps in the risk management process have concentrated on identifying, understanding, and analyzing the uncertainties that face a project that, if they occur, could have an effect on achievement of project objectives. These steps are clearly vital as it is not possible to address risks that are not identified, or are poorly understood or even misunderstood. However, the risk process cannot stop with analysis of the challenge posed by uncertainty. If nothing further is done with the information gathered in the earlier phases of the risk process, no benefit is gained by the project. Diagnosis is the not the same as cure. In most cases, simply identifying and understanding a risk does not make it go away (although this may be true in cases where the risk is epistemic, i.e., arising from ignorance or lack of awareness). Unless the risk process results in action, it is largely a waste of time.

For this reason, the next phase of the risk process is perhaps the most important in terms of enabling effective management of risk. Having identified and assessed risks, it is now necessary to decide how to respond. The effectiveness of responses will determine whether the risk exposure of the project is influenced for better or worse, resulting in increased or decreased threat and opportunity. It is therefore essential that proper attention is paid to Risk Response Planning. It is also important that this phase should concentrate equally on threats and opportunities, seeking to maximize upside and minimize downside. The aim of Risk Response

Planning is captured in the well-known song written by Johnny Mercer and Harold Arlen in 1944 and famously recorded by Bing Crosby:

> You've got to accentuate the positive, eliminate the negative ...
> Don't mess with Mister In-between,
> You've got to spread joy out to the maximum, bring gloom down to a minimum,
> Have faith, or pandemonium is liable to walk upon the scene.

Three prerequisites for effective Risk Response Planning should be in place before effort is spent on this phase:

1. List of identified and assessed risks, screened to ensure that only genuine risks remain, assessed for probability and impacts, and categorized by source of risk and area affected. Where time for response planning is limited, it will be helpful to prioritize the list of risks, so that available time can be spent on the most significant threats and opportunities first.
2. List of project stakeholders, able to act as owners of risk responses.
3. Agreed risk threshold for the project, to define the "acceptable" level of risk as a target for risk responses to meet.

If any of these prerequisites are missing, the effectiveness of response development is likely to be compromised.

- The first consideration is whether the preceding stages of the risk process have been completed satisfactorily. It is clearly essential to know which risks require responses, and which require priority attention. Outputs from the previous phases of the risk management process therefore form the key input to the Risk Response Planning phase.
- Equally important is agreement from project stakeholders that their responsibility toward the project includes a commitment to address threats and opportunities within their area of influence, taking ownership of responses where necessary. This is true at all levels of the organization, for example with the project manager taking responsibility for addressing risks at project level and senior management being prepared to develop and implement responses to strategic risks.
- Finally, the acceptability threshold is vital, to define a target against which the effectiveness of responses can be measured. Without such a target, too much effort might be spent on modifying risk

beyond what would be acceptable, or responses might not go far enough in reducing threats and enhancing opportunities.

STRATEGY BEFORE TACTICS

One weakness of the traditional approach to risk management is for the Risk Response Planning phase to focus on development of actions at a detailed level without consideration of the overall aim of responses. This can result in a lot of activity but a lack of effectiveness, especially if actions are not well coordinated. To overcome this, a two-stage approach should be followed, first defining the appropriate *strategy* for dealing with a particular risk, then designing *tactics* to implement the chosen strategy.

It is important to determine the appropriate strategy first, then to develop responses to put it into practice. This avoids the "scatter-gun" approach, where a number of alternative responses may be proposed, some of which may negate the effect of others. Determining strategy first will ensure that responses are aiming for the same goal, and should avoid nugatory effort. There is no single "best" response strategy, and each risk must be considered on its own merits. Some risks may require a combination of strategies and multiple responses, whereas others may need only one strategy with a single response.

A number of factors should be considered when deciding on the appropriate strategy to adopt. These include:

> The type and nature of the risk
> Manageability and amenability to reduction or control
> The degree of severity of impact
> Available resources
> Cost-effectiveness
> Risk-effectiveness

Having selected the appropriate strategy, attention can then be given to development of tactical responses that target individual threats or opportunities and that aim to realize the strategy.

COMMON RESPONSE STRATEGIES

A common mistake made by organizations is to pay too little attention to Risk Response Planning, thinking that "a risk identified is a risk managed." There is, however, a further significant limitation to the effectiveness of the response planning phase that is evident in the majority of cases. This is the absence of a considered response to potential opportunities facing the project. Almost without exception, Risk Response Planning as implemented on most projects concentrates exclusively on dealing with threats. Given the importance of this phase of the risk management process, it is clearly vital that equal attention should be given to opportunity responses.

In the remainder of this chapter the traditional focus of Risk Response Planning on threats is first examined, then modifications are proposed to allow development of similar approaches to dealing with opportunities.

Terminology may differ, but typical threat-based risk responses can be grouped into four basic strategies:

Avoid—seeking to eliminate uncertainty
Transfer—seeking to transfer ownership and/or liability to a third party
Mitigate—seeking to reduce the size of the risk exposure to below an acceptable threshold
Accept—recognizing residual risks and devising responses to control and monitor them

Each of these strategies aims to tackle threats in a different way, as illustrated in Figure 1. This figure takes the cause–risk–effect structure discussed in Chapter 4 (see Fig. 1) and suggests alternative ways of affecting the risk:

Avoidance aims to cut the cause–risk link making it impossible for the risk to occur, or to cut the risk–effect link so that if the risk does occur there can be no impact on project objectives.
Transfer does not change the risk directly, but involves others in its management.
Mitigation seeks to weaken the cause–risk and risk–effect linkages to make a risk less likely or less severe (or both).
Acceptance recognizes that some risks cannot be influenced, and involves taking the uncertainty and its possible impact into account within the baseline of the project.

These four strategic options are briefly discussed in turn below, together with some suggested tactics that might be adopted to implement the chosen strategy.

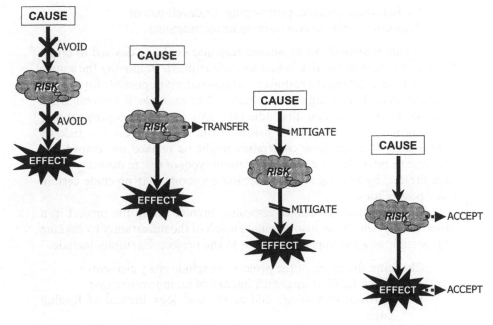

Figure 1 Threat response strategies.

Avoid

The *avoidance* strategy seeks to eliminate uncertainty and so remove a threat. It is evidently neither possible nor desirable to remove all uncertainty from a particular project, and this is not the objective of threat avoidance strategies. It may, however, be possible to eliminate particular threats, and this can be achieved by two types of avoidance response: direct and indirect. The direct avoidance responses seek to break the cause-risk link, whereas indirect responses address the risk-effect link (see Fig. 1).

Some risks arise from lack of knowledge, known technically as epistemic uncertainty (from the Greek word *episteme*, meaning knowledge). These types of threat-risk can often be tackled directly. The following actions can lead directly to elimination of uncertainty:

    Clarifying requirements
    Defining objectives
    Obtaining information
    Improving communication

Undertaking research, prototyping, or development
Acquiring expertise (via training or recruitment)

An alternative direct avoidance response might be devised to target the specific cause of the risk, where this is identified. Removing the source of the risk (or breaking the causal chain) can make it impossible for the risk to occur, thus eliminating the uncertainty. For example, if a threat arises because of the requirement to conduct the project in a developing country or using new technology, the project team may address the risk by determining whether these constraints might be relaxed or removed. It may also be possible to "design out" certain types of risk in the early stages of a project, by making strategic project decisions that preclude certain risky possibilities.

Indirect threat avoidance responses involve doing the project in a different way, which can also eliminate much of the uncertainty by making any possible adverse impact irrelevant to the project. Examples include:

Changing the scope of the project to exclude risky elements
Adopting a familiar approach instead of an innovative one
Using proven technology and/or methodology instead of leading
    edge
Building redundancy into the project design

These types of response do not remove uncertainty as such, but insulate the project from the effects of the uncertainty occurring. For example, if a project is subject to exchange-rate risk because it is costed in local currency, changing to a dollar basis would effectively eliminate the threat, since even if exchange-rate fluctuation occurs there could be no impact on the project.

Transfer

The risk *transfer* strategy aims to pass ownership and/or liability for a particular threat to another party. The ability to transfer liability for risk exposure seems attractive to many organizations, and many seek to use this strategy whenever possible. Its main use, however, is limited to financial risk exposure, since while it is possible to arrange for some other party to pay money in the event of a risk occurring, it is often difficult to enhance performance shortfalls, and it is never possible to recover lost time. It is also important to remember that risk transfer nearly always involves payment of a risk premium, and the cost must be balanced against the benefit of transferring the risk to another party.

Risk transfer responses fall into two groups, based on financial instruments and contractual arrangements.

1. Financial instruments for risk transfer include performance bonds, warranties, and guarantees, as well as more exotic arrangements such as derivatives and hedge funds. Some organizations may also consider self-insurance or use of captives (owned or rented). Traditionally financial risk transfer responses are thought to include use of insurance, where payment of a premium allows any financial penalty to be borne by the insurer, including third-party liability and professional indemnity. Strictly speaking, however, this is not a transfer strategy, since insurance effectively removes the financial risk from the project—if the risk occurs the insurer pays and there is no further cost to the project. Consequently it is more correct to view insurance as a risk avoidance response.

2. An alternative group of risk transfer responses use the contract as a means to pass liability for risk. Use of a fixed price effectively transfers financial risk to the supplier, whereas a cost-plus or reimbursable contract leaves the risk with the buyer. Other forms of contract can apportion risk in different ways between the contractual parties, including risk-reward or risk-sharing contracts, or target-cost incentivization arrangements. Specific risks can be explicitly excluded from the project through the contractual arrangements, and remain to be borne by the client or customer. Alternatively, liquidated damages or penalty/incentive payments may be agreed that pass risk to the supplier. Joint ventures, teaming, or partnership arrangements can also involve explicit risk transfer among the various parties, and this is usually captured in the contractual relationship between them.

Whichever type of risk transfer mechanism is selected, it is important to pass responsibility for management of the threat as part of the arrangement. Risk transfer not only shifts the liability for any adverse impact, but also involves a change in ownership of the risk. It must, however, be accepted that transferring the risk does not remove the threat from the project, but simply gives another party responsibility for its management. It is therefore essential that recipients of transferred risks must be able to actually manage the risks allocated to them, otherwise the project will remain exposed to an uncontrolled threat.

Mitigate

The number of threats that can be addressed by avoidance or transfer responses is usually limited. This leaves mitigation or acceptance as the strategies to be used most often. The purpose of risk *mitigation* is to reduce the "size" of the risk exposure to below a threshold of "risk acceptability." It is clearly important to define this threshold before embarking on any mitigation, since it forms the target against which response effectiveness can be measured. Acceptable risk can be determined in terms of risk severity (High/Medium/Low), or plotting regions on an iso-risk diagram or P-I Grid, or using a probability-impact ranking system (P-I Scores), as discussed in Chapter 5.

The "size" of a threat can be reduced by tackling either its probability to make it less likely, or its impact to make it less severe, or both (see Fig. 1). Preventive responses are better than curative ones, since they are more proactive, and if fully successful can lead to risk avoidance. This is illustrated by the proverb "It is better to extinguish the match than fight the fire."

- Preventive responses tackle the causes of the risk, seeking to reduce the chance of the threat occurring (i.e., lower probability). If trigger conditions for a risk can be identified, these can be targeted to make the risk less likely. (Of course, if probability is reduced to zero, then this is effectively an avoidance response.)
- Where it is not possible to reduce probability, a mitigation response might address the adverse risk impact, targeting those impact drivers that determine the extent of the severity. Early action to protect against the worst effects of a risk can make it more acceptable.

The majority of identified threats will probably be the target of risk mitigation responses, since there are not many threats where avoidance or transfer responses are possible or desirable. This type of response, however, should be very specific to the individual risk, since it addresses the particular causes of the risk and its unique effects on the project objectives.

Accept

Some threats will remain after avoidance, transfer, or mitigation responses have been taken, and others will be identified that cannot be tackled proactively within the scope of the project or the capability of the organization. These are known as residual risks. There will also be some minor threats where any response is not likely to be cost-effective compared to the possible cost of bearing the risk impact. These residual risks must be *accepted* by the project and the organization, but they also need to be proactively managed, even if they cannot be influenced in the same way as other risks. The project must recognize and accept these remaining threats, and adopt responses to protect against their occurrence.

There are two categories of acceptance response for dealing with residual threats, which can be described as "active acceptance" and "passive acceptance."

The most usual active risk acceptance response is contingency planning, including amounts of time, money, or resource to account both for known risks and for those that are currently unknown. It is useful, however, to distinguish between these two types of contingency, since one relates to defined threats (known unknowns), whereas the other deals with unforeseen risks (unknown unknowns):

- For defined risks, contingency should take the form of a risk budget, with the size related to the impact of the threat. Risk budgets should be allocated against specific threats, with agreed release conditions defining when the contingency amount should become available for use.
- Risks that are currently unforeseen must be covered by "true contingency," which reflects the amount of residual uncertainty in the project (although this may be difficult to estimate accurately, and may be no more than an educated guess).

Other more general responses can form part of a passive risk acceptance strategy to protect the project or the organization against the effects of accepted threats, including:

- Development of a risk-aware culture in the project and the organization
- Incorporating risk management into routine project processes, with regular risk reviews, reports, and updates
- Taking account of identified risks and agreed responses in project strategy, including appropriate activities in the project plan and budget

These softer responses serve to develop a robust project culture that can cope with the need to operate under conditions of uncertainty, and will allow residual threats to be accepted without disrupting the execution of the project.

Where threats with high potential impacts must be accepted, fallback plans should be developed, to be implemented in the case of the risk occurring (see below).

THE NEED FOR A PARADIGM SHIFT

It is clear that if the risk management process is to encompass management of opportunities, then the traditional approach to Risk Response Planning as described above is inadequate, since it is targeted exclusively at threats. Clearly, no one would wish to *avoid* an opportunity, nor is it usually considered appropriate to *transfer* a potential benefit to another party. *Mitigating* an opportunity to make it smaller is also the wrong approach, and simply to *accept* that an opportunity might or might not happen seems unwise.

Given that the Risk Response Planning phase has the most direct influence over risk exposure, one might expect this phase to be the part of the risk management process that most clearly targets both opportunities and threats. However, some modification is required to the standard risk response strategies to make them suitable for handling opportunities.

Since project managers and risk practitioners are used to the four common risk response strategies for threats of *avoid*, *transfer*, *mitigate*, and *accept*, it seems sensible to build on these as a foundation for developing strategies appropriate for responding to identified opportunities. This can be done by seeking to understand and generalize the underlying principle behind each threat strategy, then extending this to develop the positive equivalent approach for dealing with opportunities. The principle is illustrated in the following table, and detailed in the paragraphs below.

Threat response	Generic strategy	Opportunity response
Avoid	Eliminate uncertainty	Exploit
Transfer	Allocate ownership	Share
Mitigate	Modify exposure	Enhance
Accept	Include in baseline	Accept

Generalizing and extending the four common threat strategies results in the following concepts:

Avoidance strategies that seek to remove threats are actually aiming to *eliminate uncertainty*. The upside equivalent is to *exploit* identified opportunities—removing the uncertainty by seeking to make the opportunity definitely happen.

Risk transfer is about *allocating ownership* to enable effective management of a threat. This can be mirrored by *sharing* opportunities—passing ownership to another party best able to manage the opportunity and maximize the chance of it happening.

Mitigation seeks to *modify the degree of risk exposure*, and for threats this involves making the probability and/or impact smaller. The opportunity equivalent is to *enhance* the opportunity—increasing its probability and/or impact to maximize the benefit to the project.

The *accept* response to threats *includes the residual risk in the baseline* without special measures. Opportunities included in the baseline can similarly be accepted—adopting a reactive approach without taking explicit actions.

Each of these four opportunity strategies can be developed further, as described below.

Exploit

The aim of this risk response strategy is to eliminate the uncertainty associated with a particular upside risk. An opportunity is defined as an uncertainty that if it occurs would have a positive effect on achievement of project objectives. The *exploit* response seeks to eliminate the uncertainty by making the opportunity definitely happen. Whereas the threat equivalent strategy of *avoid* aims to reduce probability of occurrence to zero, the goal of the *exploit* strategy for opportunities is to raise the probability to 100%—in both cases the uncertainty is removed. This is the most aggressive of the response strategies, and should usually be reserved for those "golden opportunities" with high probability and potentially high positive impact that the project or organization cannot afford to miss.

In the same way that risk avoidance for threats can be achieved either directly or indirectly, there are also direct and indirect approaches for exploiting opportunities.

> *Direct responses* include making positive decisions to include an opportunity in the project scope or baseline, removing the uncertainty over whether or not it might be achieved by ensuring that the potential opportunity is definitely locked into the project, rather than leaving it to chance. For example, if an opportunity is identified that market share would be enhanced if a competitor withdrew from the market, active steps could be taken to achieve this by buying out the competitor or forming a cooperative alliance.
>
> *Indirect exploitation responses* involve doing the project in a different way to allow the opportunity to be achieved while still meeting the project objectives, for example by changing the selected methodology or technology. Where avoidance goes round a threat so that it cannot affect the project, exploitation stands in the way of the opportunity to make sure that it is not missed, in effect making it unavoidable.

Share

One common objective of the Risk Response Planning phase is to ensure that ownership of the risk response is allocated to the person or party best able to manage the risk effectively. For a threat, *transferring* it passes to another party both liability should the threat occur and responsibility for its management. Similarly, *sharing* an opportunity involves allocating ownership to another party who is best able to handle it, both in terms of maximizing the probability of occurrence, and in increasing potential benefits should the opportunity occur. In the same way that those to whom threats are transferred are liable for the negative impact should the threat occur, those who are asked to manage an opportunity should be allowed to share in its potential benefits.

Clearly it is sensible to consider project stakeholders as potential owners of this type of response, since they already have a declared vested interest in the project, and are therefore likely to be prepared to take responsibility for managing identified opportunities proactively.

A number of contractual mechanisms can be used to *transfer* threats between different parties, and similar approaches can be used for *sharing* opportunities. Risk-sharing partnerships, teams, special-purpose companies, or joint ventures can be established with the express purpose of managing opportunities. The risk-reward arrangements in such situations must ensure equitable division of the benefits arising from any opportunities that may be realized. The target-cost-incentivization type of contract is also suitable for both threats and opportunities, since it provides a mechanism for distributing either profit or loss.

It is important that risk sharing does not become mere abdication of responsibility on the part of the project manager, who should retain an active involvement in the management of all risks that could affect project objectives.

Enhance

For risks that cannot be avoided/exploited or transferred/shared, the third type of response strategy aims to modify the "size" of the risk to make it more acceptable. In the case of threats, the aim is to *mitigate* the risk to reduce probability of occurrence and/or severity of impact on project objectives. In the same way, opportunities can be *enhanced* by increasing probability and/or impact, by identifying and maximizing key risk drivers.

The probability of an opportunity occurring might be increased by seeking to facilitate or strengthen the cause of the risk, proactively targeting and reinforcing any trigger conditions that may have been identified, recognizing that if probability can be increased to 100%, then this is effectively an *exploit* response. Impact drivers that influence the extent of the positive effect can also be targeted, seeking to increase the project's susceptibility to the opportunity, and hence maximize the benefits should it occur.

Where several opportunity risks have been identified as arising from a common cause, it may be particularly cost-effective to look for generic enhancement actions that target the common cause. If these actions are successful, they will influence more than one opportunity, and could result in a significant increase in benefits to the project.

Like threat mitigations, opportunity enhancement responses are likely to be specific to the individual opportunity identified, since they address the particular causes of the risk and its unique effects on project objectives. It is therefore not possible to provide a comprehensive list of action types under this strategy, and a considerable variety of actions is to be expected.

Accept

Residual risks are those that remain after avoid/exploit, transfer/share, and mitigate/enhance responses have been exhausted. They also include those minor risks where any response is not likely to be cost-effective, as well as uncontrollable risks where positive action is not possible. The common terminology adopted for threats in these categories is to *accept* the risk, with application of contingency where appropriate, and ongoing reviews to monitor and control risk exposure.

Opportunities that cannot be actively addressed through exploiting, sharing, or enhancing can also be *accepted*, with no special measures being taken to address them. (Some suggest that the opportunity equivalent to *accept* should be *reject*, because the opportunity is not proactively pursued. The idea here, however, is that the uncertainty is being accepted, whether it has a potential for a positive or negative impact.)

In the same way as for threats, *accepting* opportunities involves taking the risk and hoping to "get lucky"—whereas for a threat this would mean hoping that the risk will not occur, for an opportunity one hopes that it will. This strategy might appear to mean taking no action at all, but a better phrase would be "Do nothing, but . . ."

One way in which opportunities can be included in the project baseline without taking special action to address them is by appropriate contingency planning. As for threats, this involves determining what actions will be taken should the opportunity occur, preparing plans to be implemented in the eventuality. For example, funds could be set aside to be spent on emerging opportunities, or resources and facilities nominated to be used if necessary.

It is also important for the project team to remain risk-aware, monitoring the status of identified opportunities alongside threats to ensure that no unexpected changes arise, and the use of an integrated risk process to manage both threats and opportunities together will assist in achieving this goal.

STRATEGY SELECTION

Given the range of possible risk response strategies, some guidance might be useful in determining the order in which strategies should be considered. Some recommend using the severity of the risk as an indicator of the preferred response strategy, as shown in Figure 2. This suggests that risks

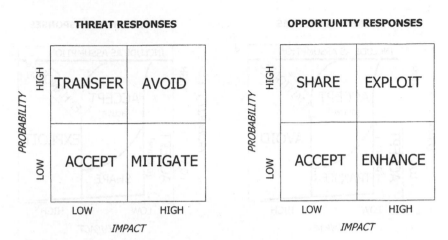

Figure 2 Selecting response strategies by risk level.

with a high probability of occurrence and high impact require aggressive treatment, and should be avoided if possible if they pose a significant threat to the project, or exploited if they represent a major opportunity. Those minor risks with low probability and low impact can be accepted, whether they are threats or opportunities, since they are unlikely to occur, and would not affect the project significantly if they did happen. High-probability/low-impact risks might be considered as safe candidates for passing to another stakeholder via transfer or share responses, since the penalty to the project would be small if the recipient failed to manage the risk effectively, either in realized negative impact or in missed benefits. Finally threats with low probability but potentially high adverse impacts should be mitigated, and opportunities with low probability but potentially high beneficial impacts should be enhanced.

This approach, however, seems to be somewhat simplistic, since cases could be made for considering different strategies for risks in each quadrant, and a more complex model might be used, for example the Risk Response Planning Charts shown in Figure 3, where the preferred strategy type is driven by the position of the risk on the iso-risk diagram.

An alternative approach might be to prioritize response strategies according to their intensity, considering first whether it is possible to select the strategy type with the biggest effect on the risk, down to the strategies with lesser effects. This is illustrated in Figure 4.

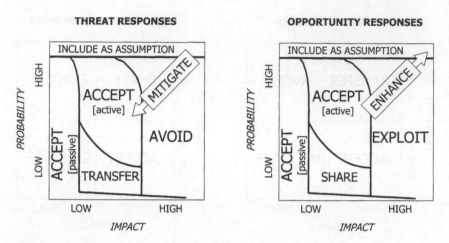

Figure 3   Risk Response Planning Charts. (From Piney, 2002.)

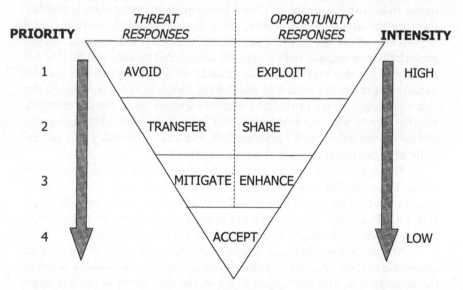

Figure 4   Selecting response strategies by intensity.

Under this approach, the more aggressive or intense strategies of avoid and exploit should be considered first, since these offer the possibility of eliminating uncertainty, by either removing a threat or capturing an opportunity. The second priority for response strategy should be to consider whether some other party might be better able to manage the risk than the project or organization itself, seeking another stakeholder who would be prepared to take the risk on, by either transferring a threat or sharing an opportunity. Where it is not considered possible or cost-effective to eliminate the risk (via avoid/exploit) or to pass ownership to another party (transfer/share), an attempt should be made to develop responses that aim to mitigate threats and enhance opportunities. The last response option is to accept residual risks that remain after these other strategies have been considered, or where other strategies are not practicable or cost-effective.

This recommended order for considering response strategies is also illustrated in Figure 4, indicating the increasing intensity of the different strategy types. Although the preferred order for considering response strategies is as shown in Figure 4, it may not be possible to implement some types of response for particular risks, requiring a different strategy to be considered. For example, a transfer strategy might be identified as the preferred response to a given threat, but the risk premium payable to execute the transfer might prove prohibitive, resulting in the need to consider mitigation or acceptance instead. Or the resources required to exploit a major opportunity may not be available in the required time frame, so a share or enhance response might be the only option.

It is also important to recognize that a single response strategy may not be sufficient to deal adequately with a particular risk, so the situation should be kept under review to determine whether the chosen response is having the intended effect. Where necessary, an alternative strategy may be required.

ALLOCATING OWNERS

Once responses have been developed, each should be assigned to an owner. This is a vital step, as the response owner will be responsible for ensuring the effective implementation of the agreed response. They will also be accountable for performing the response (or ensuring that it is performed by others). It is advisable to involve response owners in developing or refining responses that they own.

It is important to select the right owner for each risk response. This is defined as "the party best placed to manage the risk effectively." While the majority of risks may be owned by a member of the project team, any project stakeholder may be eligible to own a response. This includes other departments within the organization (for example, supplier risks may be owned by the procurement department, or resource risks by the personnel department). Some risks could be allocated to the customer, client, or users, especially performance risks or those relating to requirement uncertainty. Others may be best placed with suppliers, contractors, or subcontractors who possess specialist expertise or have responsibility for particular elements of the project. The key consideration is to determine who can make a difference to the risk.

When allocating owners, it is important to build and retain cooperation and consensus, seeking to avoid contractual wrangling or the placing of blame. The necessary resources should be provided to enable the response to be implemented, and the project manager should monitor the status of risk responses regularly, not abdicating responsibility to the response owner.

TESTING RESPONSE EFFECTIVENESS

When designing responses to risks, whether threats or opportunities, it is important to consider whether they will have the desired effect. The following seven criteria can be used to assess the effectiveness of proposed risk responses:

> *Appropriate*—the correct level of response must be determined, based on the "size" of the risk. This ranges from a crisis response where the project cannot proceed without the risk being addressed, through to a "do nothing" response for minor risks. In some cases it is entirely appropriate to stop the project until a particular risk has been dealt with, and other risks can be completely ignored. Clearly it is vital to distinguish between these two categories, so that major threats or opportunities are not ignored while the project wastes valuable time and resources on tackling minor ones.
>
> *Affordable*—the cost-effectiveness of responses must be determined, so that the amount of time, effort, and money spent on addressing the risk does not exceed the available budget or the

degree of risk exposure. (Cost-effectiveness is discussed further below.) Each risk response should have an agreed budget that is affordable within the overall project budget.

Actionable—an action window should be determined within which responses need to be completed to address the risk. Some risks require immediate action, while others can be safely left until later. It is important to identify whether action is possible in a time frame that allows the risk to be tackled effectively.

Achievable—there is no point in describing responses that are not realistically achievable or feasible, either technically or within the scope of the respondent's capability and responsibility. For example, the threat of reduced productivity rate might be tackled by a proposed response to cancel all holidays and enforce weekend working; however, if this is not possible given working terms and conditions, the response is useless. Similarly, a response to capture the opportunity of including additional functionality by immediately recruiting ten world-class engineers to the team can probably not be implemented.

Assessed—the organization needs to be confident that all proposed responses will work and be "risk-effective." This is best determined by making a speculative "postresponse risk assessment" of the risk assuming effective implementation of the response, and comparing this with the "preresponse" position. This can be achieved using an iso-risk diagram, Probability-Impact (P-I) Grid, or P-I Scoring system as discussed in Chapter 5, when the "before and after" positions can be compared. For threats the objective is to move the risk toward a lower-left-hand position on the grid with a lower P-I Score, whereas the scores for opportunities should be increased so as to move the risk toward the upper-right position. This approach can also be used to decide whether the level of planned response is sufficient, since it is possible to predict how much residual risk would remain after planned responses have been implemented.

Allocated—there should be a single point of responsibility and accountability for implementing the response. It is important to nominate a response owner for each response, who will be accountable for its implementation, although he may choose to delegate actions to others. It is recommended that each

response should have a single owner to focus this accountability.

Agreed—the consensus and commitment of stakeholders should be
obtained before agreeing responses. It is particularly important
to gain the buy-in of response owners who are expected to
implement planned actions, so that responses are not imposed
on people who are unwilling or uncommitted. As discussed
above, one way of maximizing buy-in is to involve proposed
response owners in development of the response.

Each proposed response should be tested against these seven criteria
before it is accepted for implementation, to ensure that it is likely to be
effective and achieve the intended result.

Implementing risk responses is usually not free. Each response is
likely to involve expenditure of additional time, cost, or resource. Clearly,
it is important that the organization should be prepared to spend the
required time, money, or effort in responding to identified risks, otherwise
the process will be ineffective. An important part of a risk-aware culture is
the acceptance that it is better to incur definite known cost now to avoid the
possibility of variable or unknown cost in the future. However, the
organization will require assurance that spend now is justified to remove
exposure later. It is also important to be sure that the amount of
expenditure is appropriate to the size of risk faced. For example, it would
not be wise to spend $100,000 on a response to a risk whose maximum
impact might be $10,000 (unless there were other impacts such as company
reputation, safety, or environmental implications, or "time is of the
essence" considerations).

One way of measuring the cost-effectiveness of proposed responses is
to convert the impact of the risk into money (including, for example, the
cost of delay, or the cost of rectifying performance shortfalls), and then to
calculate the Risk Reduction Leverage (RRL) factor, as follows:

$$RRL = \frac{(\text{Cost Impact})_{\text{before response}} - (\text{Cost Impact})_{\text{after response}}}{\text{Cost of response}}$$

This gives the ratio of the improvement in risk exposure to the cost of
obtaining that improvement. The larger the RRL, the more cost-effective
the response. Values of RRL less than one cost more now than they might
save later. As a guideline, effective responses should have values of RRL
above 20. RRL can also be used to compare alternative proposed

responses, allowing the most cost-effective response to be selected. Calculation of RRL is only possible, however, if all impacts of a risk can be converted into money, and if the "before and after" cost impacts can be estimated accurately.

FALLBACK PLANS AND SECONDARY RISKS

The purpose of the Risk Response Planning phase is to change the risk exposure of the project by minimizing threats and maximizing opportunities. It is important to consider what actions might be appropriate if planned responses do not have the desired or intended effect, or if the risk occurs—these actions are known as *fallback plans*. It is also important to consider the effect of planned responses on the risk profile of the project, since these actions may in themselves directly introduce new uncertainty, known as *secondary risks*.

Fallback Plans

For risks with potentially major impacts (both threats and opportunities), it is advisable to develop *fallback plans* ready for implementation should the risk occur. This is analogous to preparing disaster recovery plans or business continuity plans. A fallback plan should be fully defined, planned, costed, and resourced. It should also have defined unambiguous trigger conditions, which determine when the risk has occurred and therefore when the fallback plan is to be implemented. The aim of a fallback plan for a threat is to minimize the adverse impact of the risk, to prevent knock-on effects into other areas of the project, and to restore control. Fallback plans for opportunities aim to maximize the unexpected advantage.

Secondary Risks

Whenever a risk response is implemented it will inevitably change the risk profile of the project. Clearly the response is designed to improve the situation, but it cannot be assumed that it will work as planned, or indeed whether the planned outcome is the only effect a response might have. The law of unintended consequences can apply to risk responses in the same way as elsewhere. Sometimes implementation of a response may introduce more risk into the project than it removes.

Risks that arise as a direct result of implementing a response are termed *secondary risks*. These should be identified for responses to key risks, and secondary risks should be assessed in the same way as primary risks. The project team should determine whether the risk position after implementation of a response will be better or worse than it was beforehand. For example, suppose that Response R is a mitigation action that has been proposed to address a primary threat (Risk A), with the result that the original Risk A is reduced to Risk A*. However, if Response R introduces a new Secondary Risk S, the project team needs to test whether $A* + S < A$, as illustrated in Figure 5.

For example, the risks associated with driving to an appointment in an unfamiliar city (car might break down, may get lost, unable to find parking, traffic delays, etc.) can be avoided by taking the train. This response introduces a new set of risks (train cancellation, unexpected delay, no transport from station to appointment, etc.). The secondary risks associated with choosing to travel by train should then be compared with the uncertainties of driving to determine whether to implement the response.

An unfortunate example of secondary risk occurred when a risk assessment for the refurbishment of a chemical reactor identified a risk that

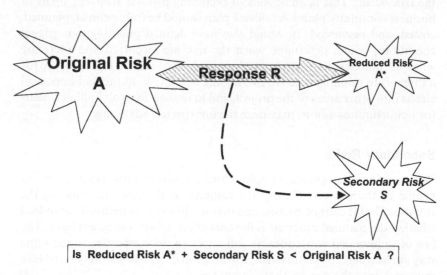

Figure 5   Secondary risk.

in-house specialist welders might not be available when required. The
chosen response was to arrange a call-off contract with another company
to supply replacement welders at short notice if needed. When the project
was underway, the entire small team of in-house welders became ill at the
precise time that they were required, so the risk response was implemented
and the replacement staff were brought on-site. However, they lacked the
necessary union accreditation and were refused entry by the site foreman,
who called a strike in protest at management's attempt to use nonunion
labour, leading to a lengthy delay. The response to the original threat had
introduced a new secondary risk (replacement welders might not be
acceptable), which had not been identified or managed, and which
occurred seriously affecting the project.

The position of secondary risks is quite clear when considering
threats. The aim of a response is to deal with the original threat without
introducing a new threat, or at least to ensure that the closing postresponse
position is no worse than the situation preresponse. When opportunities
are included in the picture, however, analysis of secondary risks becomes
rather more complex.

A number of secondary-risk scenarios might be envisaged, since each
original risk and secondary risk could be either a threat or an opportunity.
In other words, the "unintended consequences" from a particular response
might be either positive or negative:

- The worst case is where a response to a threat introduces another
  secondary threat, perhaps making the situation worse than it was
  before the response was implemented. This is the standard view of
  secondary risks, with the potential to generate a "vicious circle" of
  increasing negative risk, as shown in Figure 6.
- The comparable upside case is when a planned response to deal
  with an opportunity may have a double benefit and introduce a
  secondary additional opportunity, which would be the best case
  possible, producing a "virtuous circle" of increasing upside (see
  Fig. 7).
- It is also possible that a response designed to address a threat not
  only achieves its aim, but also introduces an additional upside; i.e.,
  the secondary risk is an opportunity. In this case there might be
  two beneficial results, with the original threat being reduced or
  eliminated and a new opportunity being created.
- Similarly, when a response is planned to tackle an opportunity, it
  may result in a secondary threat. Here the potential benefit of

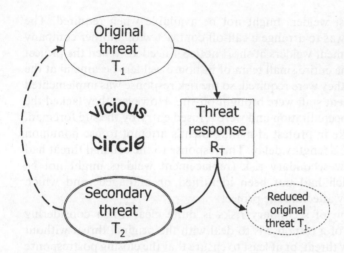

Figure 6   Vicious circle of threat and secondary threat.

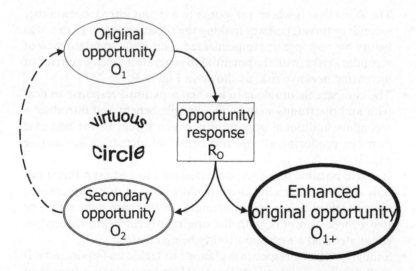

Figure 7   Virtuous circle of opportunity and secondary opportunity.

dealing with the opportunity might be discounted by the introduction of a new threat, and the balance must be considered carefully.

These last two cases might be combined as shown in Figure 8, where the best possible outcome is to execute a response (R_T) to deal with the original threat (T_1) and at the same time introduce a secondary opportunity (O_2). In the same way, the situation to avoid is where a planned response (R_O) to the original opportunity (O_1) results in a new secondary threat (T_2).

The "car or train" example above provides a simple illustration of the double-benefit secondary risk (where a response to a threat creates a new opportunity), since choosing to take the train creates an opportunity to work or sleep while traveling, as well as avoiding the car-related threats. Considering secondary opportunities as well as threats can alter the decision about whether to implement the response.

A similar situation occurred when a pharmaceutical company identified a number of threats arising from the logistical complexity of a clinical trial protocol, including possible problems in recruiting investigators or patients and poor data quality. The selected response to optimize the pro-

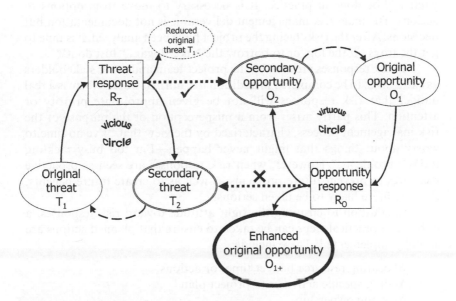

Figure 8 Combined vicious/virtuous circles.

tocol led to opportunities to reduce the cost of the trial and enhance its scope.

MAKING IT HAPPEN

If the Risk Response Planning phase is the part of the risk management process where a difference can be made to risk exposure, this phase needs to be done well. It is, however, not enough simply to think about what can be done. Nothing will actually change until planned actions are implemented. A well-known riddle illustrates this point: "There are five frogs sitting on a log, and four decide to jump off. How many frogs are on the log?" The correct answer is five, since there is a difference between deciding and doing. Having decided to jump, the frogs remain on the log until they actually do jump.

Too many organizations fail to make the transition from planning to implementation in their execution of the risk management process. For these, the end deliverable is a set of documented risks and perhaps some analysis outputs. The Risk Response Planning phase does need to consider what *can* be done to address identified risks, but it also must go on to decide what *will* be done in practice. It is necessary to move from options to actions. The main risk management deliverable is not documentation but decisions. After the risks facing the project have been analyzed, it is time to get the frogs off the log, or to borrow the Nike slogan, "Just do it!"

If risk responses are seen by the project team and other stakeholders as extra tasks to be completed in addition to "normal work," there is a real danger that risk responses will not be given appropriate priority or attention. This partly arises from a misperception of the purpose of the risk management process, characterized by the view that "I've no time to worry about things that might never happen—I'm too busy tackling today's problems." However, when risk responses are seen as proactive measures to prevent future problems as well as to create future benefits, they are more likely to be taken seriously.

In addition to ensuring the right attitude toward risk responses, a number of practical steps can be taken to ensure that planned actions are actually implemented. These include:

Allocating resource/budget/time for actions
Adding specific activities to project plan
Agreeing ownership
Monitoring, reporting, and reviewing responses

Taking these simple steps results in risk responses being treated in the same way as "ordinary project tasks," with a plan and budget, a responsible owner, and the requirement to report progress. In this way the project team and other stakeholders will regard implementation of risk responses as a part of their job, rather than something additional to be resisted.

Risk Response Planning is an essential part of the risk process, since this is where effective actions can be defined to deal appropriately with identified risks. Unfortunately this phase of the process as currently implemented in most organizations totally fails to take account of the existence of opportunities, concentrating instead on strategies to deal with threats. However, a simple extension of the familiar threat strategies as described in this chapter produces a comparable set of response types that can address opportunities in similar ways. This gives the project manager a comprehensive range of weapons with which to tackle the risks facing the project, whether they be threats or opportunities. Without such responses, it is impossible to manage risk effectively. Specific actions focused around targeted strategies will, however, ensure that threats are minimized and opportunities maximized—as long as agreed responses are actually implemented in practice. The final chapter in this part of the book addresses how to keep the risk process on track to achieve these aims.

REFERENCES

Ben-David, I., Raz, T. (1999). An integrated approach to risk response development in project planning. Working Paper 6/99. Tel Aviv, Israel: Israel Institute of Business Research.

Feldman, M. L., Spratt, M. F. (1999). *Five Frogs on a Log: A CEO's Field Guide to Accelerating the Transition in Mergers, Acquisitions and Out Wrenching Change*. New York: Harper Business.

Hillson, D. A. (1999). Developing effective risk responses. In: *Proceedings of the 30th Annual Project Management Institute Seminars, Symposium*, presented in Philadelphia, USA, 11–13 October, 1999.

Hillson, D. A. (1999). Take no risks with risk. *Project* 12(1):14–16.

Hillson, D. A. (2001). Effective strategies for exploiting opportunities. In: *Proceedings of the 32th Annual Project Management Institute Seminars, Symposium* (PMI 2001), presented in Nashville, USA, 5–7 November, 2001.

Institution of Civil Engineers (ICE) and Faculty, Institute of Actuaries. (1997). *Risk Analysis, Management for Projects (RAMP)*. London: Thomas Telford.

Nordland, O. (1999). A discussion of risk tolerance principles. *Safety Critical Systems Club Newsletter* 8(3):1–4.

Piney, C. (2002). Risk response planning: selecting the right strategy. In: *Proceedings of the 4th European Project Management Conference* (PMI Europe 2001), presented in London, UK, 6–7 June 2001.

Project Management Institute. (2002). *A Guide to the Project Management Body of Knowledge (PMBoK®)*. 2000 ed. Philadelphia: Project Management Institute.

Simon, P. W., Hillson, D. A., Newland, K. E., eds. (1997). *Project Risk Analysis, Management (PRAM) Guide*. High Wycombe, Bucks, UK: APM Group.

UK Office of Government Commerce (OGC). (2002). *Management of Risk—Guidance for Practitioners*. London: Stationery Office.

8
Monitoring, Control, and Review

The risk process described so far in Chapters 3–7 (and summarized in Fig. 2 of Chapter 2) has set objectives for risk management on a particular project (Definition), then identified uncertainties that, if they occurred, would have a positive or negative effect on one or more objectives (Risk Identification). Those risks have been described, grouped, and prioritized through qualitative Risk Assessment, and their combined effect on the project may also have been addressed using quantitative Risk Analysis, prior to developing appropriate strategies and actions for dealing with the uncertainty (Risk Response Planning). The aim throughout is to implement a single risk management process that allows the organization to address appropriately the threats and opportunities to which the project is exposed, minimizing possible downside impact and maximizing the upside, while maintaining focus on achievement of project objectives.

At this point in the process it is interesting to consider how the risk exposure of the project might have changed. In reality, the answer is likely to be "very little." A great deal of risk-related data has probably been generated, and a lot of time may have been spent thinking about risk, but the risk process from Definition to Risk Response Planning does not significantly alter the risks faced by the project. After these phases have been implemented, much more is known about the range of threats and opportunities, and which are the important ones, and what can be done about them. But the process so far has concentrated on *analysis*. Risk exposure remains unchanged until *action* has been taken—to return to the closing imagery of the previous chapter, it is necessary to get the frogs off the log. Without action, risk management is just an exercise in satisfying intellectual curiosity, thinking hypothetically about things that might or might not happen, and speculating about what might be done. One of the most frustrating comments that can be made to a project manager facing a crisis or missed opportunity is for someone to say "I could have told you that would happen!" Although there are some cases where simply exposing an uncertainty can result in its disappearance, merely knowing about a risk is not usually enough to change the likelihood of it happening or its impact if it does occur.

This is why the final phase in the risk process (Risk Monitoring, Control, and Review) is so important, to the point where some standards and methodologies actually call this step "risk management." This is where planned actions are implemented to achieve the intended changes to risk exposure, minimizing and avoiding threats while maximizing and exploiting opportunities, to increase the confidence that project objectives can be achieved. The effectiveness of the risk management process (and particularly of risk responses) is monitored and reviewed, and changes in the level of risk faced by the project are assessed, to ensure that the expected benefits are being reaped.

For the typical risk management process, it is the Risk Monitoring, Control, and Review stage that results in actual reduction of threats. Similarly for the risk process, which has been broadened to include opportunity, this is the stage where the upside is actually gained. This chapter describes the steps that a risk process should include to achieve the potential, and highlights changes that might be required to ensure that opportunity enhancement occurs alongside threat reduction.

COMMUNICATING RESULTS

Albert Einstein (1879–1955) reputedly said, "The major problem in communication is the illusion that it has occurred." Communication is recognized as a key element of project management, and one that is often a weak area in terms of both personal skills and organizational structures. In the same way that communication is a critical success factor for project management, it is also a vital ingredient in effective risk management. A risk process can result in a comprehensive set of identified risks, objectively prioritized, and each with carefully targeted action plans. But if the project manager is the only person who knows about them, it is unlikely that very much will happen.

It is important, therefore, to communicate the results of the risk process to project stakeholders, and the key is to identify those stakeholders and their information needs. The outputs of the project risk-reporting process can then be designed to meet those needs specifically and precisely. A structured approach should be followed, either as part of overall project communications planning, or as a specific exercise for risk reporting, for example:

Identify Stakeholders

A stakeholder is defined as any person or party with an interest in the project. By implication, most stakeholders are also able to influence the project to some degree. All stakeholders should be defined, together with their level of interest and degree of influence. A typical list of project stakeholders might include some or all of the following:

Project sponsor
Project manager
Project team members
Project risk specialist or analyst
Senior management
Other company departments
Clients
Users
Customers
Subcontractors
Suppliers
Regulatory authorities
General public

Identify Stakeholder Risk Information Needs

Requirements for risk management information should be determined for each group of stakeholders, linked to their interest in the project and what "stake" they hold in its success. Identifying risk communication requirements includes answering the following questions:

What risk information is required?
How will it be used?

What level of detail and precision is required?
When must risk information be supplied?
What time delay is acceptable (if any)?
How frequently are updates needed?
How should information be delivered?

This information should be recorded in a stakeholder risk information needs analysis, such as the example shown in Table 1.

Design and Validate Communications

Based on the stakeholder risk information needs analysis, outputs from the risk management process should be designed to meet the identified requirements. This should include consideration of content and delivery method, as well as defining responsibilities:

Content: A range of risk reports may be required at different levels of detail, delivered at different frequencies. It is more efficient to design reports in a hierarchical manner if possible, to allow high-level reports to be generated as summaries of low-level reports, to avoid the overhead of producing multiple outputs.

Delivery method: Alternative methods of delivering risk management outputs to stakeholders should be identified, and the appropriate means to be used for each. These might include written reports in hard-copy or electronic format (e-mail, intranet, website, accessible databases), verbal reports (briefings, presentations, progress meetings), graphic or numerical outputs (tables, charts, posters), etc.

Responsibilities: Each output should have a defined owner responsible for its production, and a designated approval authority. It may also be helpful to identify those whose contributions will be required, and who will receive the output for information. A RACI analysis might be used to document this (Responsible, Approval, Contributor, Information).

Proposed risk outputs should be documented as shown in Table 2. Information in this table should be cross-compared with Table 1 to ensure that all stakeholder requirements are met by the proposed outputs. The design process should also include a validation step, checking with each group of stakeholders that the proposed communication meets their need.

Table 1 Stakeholder Risk Information Needs Analysis (with sample entry)

Stakeholder	Interest in project	Information required	Purpose	Frequency	Format
Project Sponsor	Meet business case	Project status summary Risk status summary Problems outside project control	Project monitoring Manage key risks Assist project manager	Monthly Monthly Immediate	2-page hard-copy report 2-page hard-copy report Verbal plus e-mail

Table 2 Risk Outputs Definition (with sample entry)

Title	Content	Distribution	Frequency	Format	Responsibility
Risk status summary	Summary of project status Summary of risk status Key risks Changes since last report Recommended actions	Project Sponsor Project Manager Program Review Board Corporate Risk Manager	Monthly	Two-page hard-copy text report, including "Top risks" table, risk trend graph, and Action table.	Project Manager (with input from key team members)

Implement and Review

The agreed risk communication strategy and outputs should be documented, either in the project Communications Management Plan or as part of the Risk Management Plan (see Chapter 3). The communications system should then be implemented, delivering agreed outputs to stakeholders as defined in the plan. After one or two cycles of risk reporting, the communications process should be reviewed with key stakeholders to check whether their needs are being met, or whether adjustments are required. Periodic reviews of the risk communications process should also be planned during the life of the project, as the risk information needs of stakeholders are likely to change as the project progresses.

Each project stakeholder has a different requirement for risk-related information, and the risk process should recognize this and deliver timely and accurate information at an appropriate level of detail to support the needs of each stakeholder. It is not adequate to simply take raw risk data

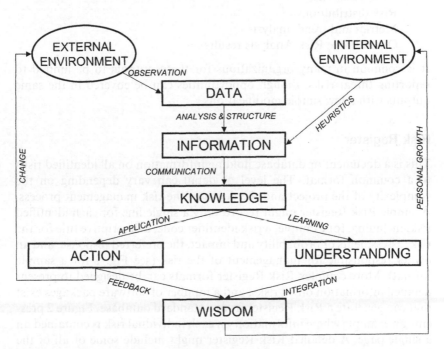

Figure 1 Data, information, knowledge, and wisdom.

and pass it on or make it available to the project stakeholders. *Data* needs to be analyzed and structured to transform it into *information*, and then actively communicated to create *knowledge,* which can form the basis for *action*, as illustrated in Figure 1 (which also shows the transformations necessary to generate *understanding* and *wisdom*, as well as the feedback loops to the internal and external environments).

TYPICAL RISK PROCESS OUTPUTS

A number of key qualitative and quantitative outputs are likely to be required as deliverables from the risk process, each of which needs to address both threats and opportunities in an equitable manner. Typical outputs are addressed in turn below, including:

> Risk Register
> Summary Risk List
> "Top risks" list
> Risk distributions
> Metrics and trend analysis
> Quantitative Risk Analysis results

It is common in many organizations for these outputs to be limited to reporting threat risks, though opportunities can be covered in the same outputs with a few simple modifications.

Risk Register

This is a document or database holding information on all identified risks in a common format. The level of detail can vary depending on the complexity of the project and the depth of the risk management process. A simple Risk Register might record only a single line for each identified risk, including, for example, a risk identifier code, a descriptive title for the risk, assessments of probability and impact, the preferred response, and an owner responsible for management of the risk (see Fig. 2 for a sample format). More complex Risk Register formats can be designed to present detailed information on all risks, and a number of software packages exist that can generate a Risk Register from a standard database. Figure 2 presents an example where information on each individual risk is contained on a single page. A detailed Risk Register might include some or all of the fields listed in Table 3.

Project Number:			Client:	
Project Title:			Project Manager:	
Risk Ref.	RBS Ref.	WBS Ref.	Risk Owner:	
Risk Type: (T/O)	Risk Status: (Draft/Active/Closed/Deleted/Expired/Occurred)			
Risk Title:				
Risk Description:				

Cause of Risk	Effect on Objectives		
	Objective	Impact Rating NII/VLO/LO/MED/HI/VHI	Impact Description
	Time		
	Cost		
	Performance		
Probability Rating Nil/VLO/LO/MED/ HI/VHI	Other		
Date Risk Raised:	Date Risk Closed/Deleted/Expired/Occurred:		

Risk Response-Preferred Strategy :			
Action(s) to implement strategy	Action Owner	Action by Date	Status
Comment/Status :			

Figuro 2 Risk Register oamplo format.

Table 3   Risk Register Fields

Project data (project title, client, contract reference number, project manager, start/end dates, etc.)

Unique risk identifier, perhaps based on RBS reference (see Fig. 5 of Chapter 5 for a sample RBS)

Risk type: threat or opportunity (this information can be incorporated into the risk identifier, for example using a T prefix for threats or O for opportunities)

Short risk title

Risk description, detailing the uncertainty itself

Cause description, including background information and trigger conditions

Impact description, relating the risk to one or more specific project objectives

RBS area (generic source of risk)

WBS areas (parts of project affected)

Date identified

Name of person identifying risk

Probability of occurrence (one of Very High/High/Medium/Low/Very Low)

Impact on project duration (Very High/High/Medium/Low/Very Low)—negative for threats, positive for opportunities

Impact on project budget (Very High/High/Medium/Low/Very Low)—negative for threats, positive for opportunities

Impact on project performance (Very High/High/Medium/Low/Very Low)—negative for threats, positive for opportunities

P-I Scores, calculated automatically based on assessments of probability and impacts

Current risk ranking, based on P-I Scores

Impact window (start/end dates when risk might affect project if it occurred)

Risk owner (responsible for overall management of the risk)

Preferred response strategy (avoid, exploit, transfer, share, reduce, enhance, accept)

Detailed risk response actions, each with action owner, action window, success criteria, budget, current status, next step, etc.

Secondary risks (if any) associated with planned responses

Postresponse assessment of risk (i.e., expected probability and impacts if the planned responses are effective as intended, possibly including modified P-I Scores)

Risk status (e.g., draft, active, closed, deleted, occurred, etc.)

Data valid date

Next review date

Historical narrative, detailing completed actions and subsequent status changes on the risk since it was identified

---

Clearly these data elements are equally relevant for describing both threats and opportunities, and can be recorded for both, though there are some differences. For example, descriptions of impacts will be negative for a threat (delay, extra cost, performance shortfall, etc.), whereas impacts for an opportunity will be positive (saving time or cost, enhancing performance, etc.). Different response strategies also apply to threats and opportunities (as described in Chapter 7).

Some organizations may choose to present a separate Threat Register and Opportunity Register, though this runs counter to the concept of implementing a combined approach to both types of uncertainty. In principle, since the key characteristics of all risks are the same whether threats or opportunities, they can both be described within a single Risk Register. It may, however, be prudent to use a numbering scheme for the risk identifier code that indicates which type of risk is being described, for example using a prefix of T for threat and O for opportunity.

Summary Risk List

This is a brief listing of risks identified for a project, usually limited to those that are currently active (i.e., excluding risks that have either occurred or timed-out or been resolved), allowing the full list of risks to be presented in a single table. The Summary Risk List can be ordered in various ways, including by RBS, WBS, risk ranking, risk owner, action window, etc. An example is shown in Figure 3. It is recommended that threats and opportunities should both be included in the Summary Risk List, allowing the overall risk picture to be seen, though they need to be distinguished using some coding scheme as for the Risk Register.

"Top Risks" List

This is a prioritized risk list showing the worst threats and best opportunities, usually ranked by P-I Score. Many organizations present the "Top Ten," though this may introduce a false threshold to limit the number of "top" risks artificially to ten—the number on the list should reflect those risks requiring urgent or focused risk management attention. This list may be presented with both threats and opportunities integrated together, or as two sublists indicating top threats and top opportunities separately. The "Top risks" list is used to concentrate action on the most significant and important risks, so it is most useful when it combines threat and opportunity information into a single list.

Project Number:	PX42651/SC3		Client:	MegaCorp Inc			Review date: 1 April 2004	
Project Title:	ASCENT System Upgrade		Project Manager:	Seymour Goodnews				

Risk Ref.	RBS Ref.	WBS Ref.	Risk Title and Description	Cause	Impact	Probability (L,M,H)	Impact on Time (L,M,H)	Impact on Cost (L,M,H)	Impact on Performance (L,M,H)	Action(s)	Action Owner
T00175	1.2	3.4.1	Requirement volatility. Although the plan assumes that there are no significant changes in requirements, the client may request or require change.	Client has not defined user spec.	Rework to design documentation, or repeat work if later in project.	M	L	L	H	Ensure system is as generic as possible to allow for likely level of changes.	System Design Manager

Figure 3 Summary Risk List sample format.

Risk Distributions

Useful management information can be generated from the distribution of identified risks across various categories, as analyzed during the qualitative Risk Assessment phase (discussed in Chapter 5). The following categorizations might form part of the risk information reported to stakeholders—risk priority, Risk Breakdown Structure (RBS), Work Breakdown Structure (WBS), Organizational Breakdown Structure (OBS), or time windows. These are outlined below:

Risk Priority

The Probability-Impact Grids illustrated in Figures 1, 2, and 3 of Chapter 5 allow identified risks to be grouped by the two dimensions of probability of occurrence and level of impact, leading to categorization for prioritization purposes. For example, risks can be divided into red/amber/green or high/medium/low-priority groups. Risk reporting outputs might include the numbers of risks in each priority category, or perhaps the total P-I Score for the risks in each category (see Fig. 4 of Chapter 5). The distribution of risks across these categories gives an indication of the overall level of risk faced by the project, and monitoring the movement of risks between the groups allows trends in risk exposure to be tracked. This works equally well for both threats and opportunities, though the aim is to move threats from red/high to green/low, whereas it is desirable to improve green/low opportunities toward the red/high zone. Graphic outputs are well suited to illustrating this type of categorization information. A simple example is presented in Figure 4, showing changes in numbers of threats and opportunities between one risk assessment and the next. The preferred distribution is high opportunity/low threat.

Risk Breakdown Structure (RBS) Category

If an RBS was used for risk identification and assessment as described in Chapters 4 and 5, distributions of identified risks can be drawn up across the various RBS categories. As for risk priority groupings, these can show either total number of risks in each RBS category or total P-I Score. The aim of such a distribution is to reveal common sources of risk for the particular project, and this can be analyzed at different levels of the RBS hierarchy from generic high-level types of risk to more detailed causes. It is probably best in this case to separate out threats and opportunities, to allow the distribution of each to be seen clearly. Management attention can

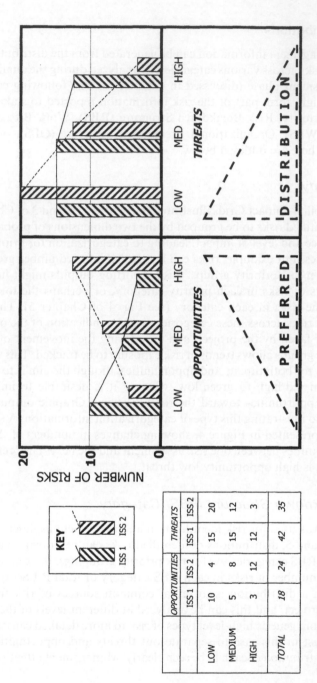

Figure 4 Risk distribution by priority category.

then be focused on those risk sources that give rise to the greatest risk, either major sources of threat or significant sources of opportunity.

Work Breakdown Structure (WBS) Category

Identified risks can also be mapped into the project WBS to expose hot spots of risk on the project, namely, those parts of the project with the highest level of associated risk. As for RBS categorization, this either can plot the number of identified risks against each WBS element, or can show the total P-I Score. A graphic showing combined RBS/WBS distribution of numbers of risks is shown in Figure 6 of Chapter 5, indicating the combination of sources of risk and areas affected. The complexity of the output suggests that separate graphic representations of threat distribution and opportunity distribution will probably present the information more clearly than attempting to combine these into a single picture. This will reveal the areas of the project under greatest threat, as well as those parts containing the highest level of opportunity, allowing management attention to be focused for maximum efficiency and effect.

Organizational Breakdown Structure (OBS) Category

As for RBS and WBS, identified risks can be categorized by risk owner, using the project OBS if available, indicating the person or party best able to manage the risk. Preliminary allocation of risk owners in this way can aid development of risk responses by early involvement of the responsible party in defining actions that are appropriate, feasible, effective, etc. The reporting of risk distribution by OBS can also indicate to the project manager when particular team members may need assistance or support, either because they have a significant risk exposure in the part of the project for which they are responsible, or if the additional risk management tasks might distract them from other important project work.

Again as for RBS and WBS distributions, it is possible to report the combined position of both threats and opportunities in a single risk ownership output, or these may be presented separately.

Timing Information

A useful dimension for risk reporting is to show timing information, including both "impact windows" (when the risk might affect the project if it were to occur) and "action windows" (when action is possible and likely to be effective). Clearly there should be no overlap between the two windows for a given risk, as this would suggest that actions might not be completed before the risk could occur. A graphical representation of impact and action windows would show if this situation could arise. It would also indicate periods in the project when risk response activity might be highest, which might have implications for project resourcing. Finally, such an output would show the time periods when the project is most at risk from the potential effects of uncertainty (either positive or negative), assisting decisions in allocation of contingency to the project (cost, time, resource, performance, etc.), and perhaps indicating the need to consider scheduling strategies such as parallelism or crashing.

Metrics and Trend Analysis

Several alternative metrics can be defined to monitor changes in risk exposure, either in the whole project or in various subcategories as defined above. These metrics often provide only a gross indication of risk trends, but can be useful in assessing the effectiveness of the risk management process. The four most common metrics are:

> Number of active risks
> Number of closed risks
> Total P-I Score for active risks
> Average P-I Score for active risks

The last two of these can be combined into a measure known as the Relative Risk Exposure Index. The idea is that true risk exposure is more than just the total "size" of the risks that face the project, but it also needs to take account of their "weight." Thus a project might be more at risk with twenty small risks than with three large ones. Relative Risk Exposure Index (I_{RR}) accounts for both number and severity, and is calculated by taking the product of Total P-I Score (T) and Average P-I Score (A) at the current point in time ($T_c * A_c$) and dividing it by the baseline value ($T_b * A_b$). Thus:

$$I_{RR} = \frac{T_c * A_c}{T_b * A_b}$$

The resulting value of I_{RR} is 1.0 (one) if the current level of risk exposure is the same as the baseline, with less than one indicating reduced risk exposure compared to the baseline, and higher than one meaning increased risk exposure over the baseline.

This only works, however, if all the risks included in the calculation are of the same type, and I_{RR} is usually calculated for threats only. When calculating P-I Scores for threats and opportunities, a high P-I Score for a threat is unwelcome since this represents the possibility of a serious negative effect. Conversely, high P-I Score for an opportunity is a good thing, since this means the potential for bigger benefits. Clearly, a mixture of threats and opportunities cannot be combined in the calculation of I_{RR}, as it would be impossible to interpret the result. It then becomes necessary to calculate two values for I_{RR}, one for threats and one for opportunities. In this case, changes in $I_{RR\text{-}THREAT}$ can be interpreted as usual, with higher values meaning worsening exposure to threats compared to the baseline situation. Changes in $I_{RR\text{-}OPPORTUNITY}$ have a different meaning, since higher values mean increased opportunity.

Having defined various metrics for risk exposure and measured them as the project proceeds, it is then possible to perform trend analysis on the results to provide useful management information on the way risk exposure is changing on the project. This might also indicate the need for modifications to the risk management process based on the direction and strength of trends.

Examples of risk metrics and trends are given in Table 4, with separate figures for threats and opportunities. The data in this table are presented graphically in Figure 5. This shows that the trend in overall threat exposure ($I_{RR\text{-}THREAT}$) follows the change in number of active threats since the average P-I Score remains nearly constant, whereas the shape is different for opportunity exposure ($I_{RR\text{-}OPP}$). The main difference is between Issue 3 and Issue 4 of the risk assessment, when threat exposure is seen to reduce from above 1 to below 1 (a good thing), but the opportunity index halves from nearly 6 to below 3 (a bad thing). This result may suggest a need to refocus the risk process on opportunity management. The closing position at Issue 5 shows threat exposure about half that at Issue 1, while opportunity exposure is about four times higher.

Despite the attraction of numerical metrics, the temptation to over-interpret must be resisted, since these figures are manipulations based on values that involve several levels of judgment. For example, I_{RR} uses the average and total P-I Scores, which depend on the efficiency of risk identification (how many identifiable risks were actually identified) and the objectivity of risk assessment (how their probabilities and impacts were perceived). Changes in these base values could have a significant influence on the calculated value of the metric, with unpredictable results.

Table 4 Sample Risk Metrics Data

	Threat data					Opportunity data				
	Number of active threats (N_t)	Number of closed threats	Total Threat P-I Score (T_t)	Average Threat P-I Score ($A_t = T_t/N_t$)	Relative Risk Exp Index $I_{RR\text{-}THREAT}$	Number of active opps (N_o)	Number of closed opportunities	Total Opp P-I Score (T_o)	Average Opp P-I Score ($A_o = T_o/N_o$)	Relative Risk Exp Index $I_{RR\text{-}OPP}$
Issue 1	20	0	15.5	0.8	1.00	5	0	2.3	0.5	1.00
Issue 2	22	5	18.2	0.8	1.25	8	0	4.8	0.6	2.72
Issue 3	30	12	21.7	0.7	1.31	9	2	7.5	0.8	5.91
Issue 4	23	23	16.1	0.7	0.94	13	4	6.2	0.5	2.79
Issue 5	14	35	9.3	0.7	0.51	6	11	5.0	0.8	3.94

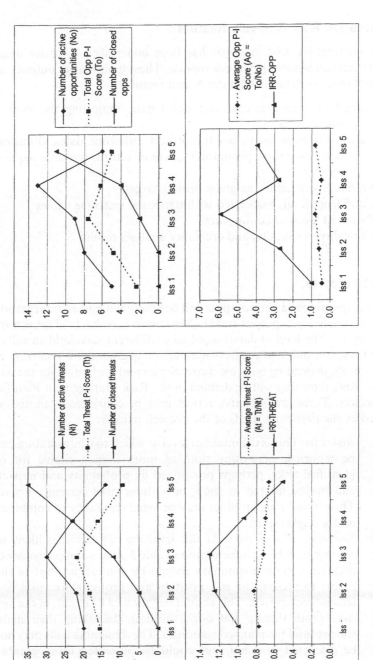

Figure 5 Risk metrics and trends.

Quantitative Risk Analysis Results

Where quantitative Risk Analysis has been undertaken there are many outputs that can be included in risk reports. These quantitative outputs are described in more detail in Chapter 6, and include:

> Spread between best case and worst case, indicating the level of uncertainty
>
> Calculated expected values for project milestone dates, duration, cost, cash flow, profitability, rate of return, resource requirement, etc.
>
> Probability of achieving given project targets
>
> Confidence levels associated with particular outcome values
>
> Required levels of contingency
>
> Criticality, sensitivity, and cruciality analysis results

Report Formats

It is likely that these various outputs will be combined into an information package for delivery to project stakeholders, most commonly as some form of risk report. The level of detail supplied to different stakeholders will of course vary, but different projects may also implement risk reporting at various levels, depending on project complexity and the budget for the risk process. Risk reporting will be defined in the Risk Management Plan for each project. Three levels of risk report may be considered, tailored as required by the particular needs of the project, namely:

1. *Risk List*. This is the minimal level at which risk information can be reported, presenting nothing more than a simple list of identified risks, perhaps prioritized by probability and impact, and maybe filtered to show only those risks currently active. Threats and opportunities may be listed separately or combined into a single report.

2. *Summary Risk Report*. At this level the risk report is likely to contain only basic information (Table 5 gives a sample contents list). An executive summary should be used at the start of the report to present the risk position and project status in a few concise paragraphs. This should be followed by details of the key risks (both threats and opportunities), describing their main features and the planned responses. The remaining risks may not be discussed in detail, but could be presented using graphic

Table 5 Summary Risk Report Sample Contents
List

Executive Summary
Project Status and Overall Risk Status (at this review)
Top Risks with agreed actions and owners
Key Changes since last review
Risks outside project scope or control
Conclusions and Recommendations
Appendix: Risk Register

---

analyses and pictures. Key changes since the last review should be summarized. Risks outside the scope of the project may also be listed, to draw them to the attention of senior management for action. The report should end with conclusions and recommendations, briefly assessing the current situation and summarizing those actions that are required to keep risk exposure within

Table 6 Detailed Risk Report Sample Contents List

Executive Summary
Project Status and Overall Risk Status (at this review)
Top Risks with agreed actions and owners
Risk Distributions
  High/Medium/Low risks
  Causal analysis (mapped to RBS)
  Effects analysis (mapped to WBS)
Quantitative Analysis [where used]
  Model inputs and structure
  Key outputs and analysis
Changes since last review, including metrics and trend analysis
Risks outside project scope or control
Conclusions and Recommendations
Appendices
  Risk Register
  Risk Data Sheets
  Quantitative analysis data [where used]
  Other detailed supporting data

---

acceptable limits. The Risk Register or Summary Risk List may
be supplied as an appendix to the Summary Risk Report.

3. *Detailed Risk Report.* This type of report will contain full details
and analysis of all identified risks, with supporting information
(see Table 6 for a sample contents list). The report should start
with a brief executive summary, then summarize the project current status and overall risk position. A review of current risks
may follow, highlighting the top threats and opportunities. A
detailed analysis of individual risks may be presented, covering
changes in the status of identified risks, and listing any new risks
identified since the last report. Risk distribution data can be
presented and discussed, identifying common sources of risk, hot
spots of exposure in the project, etc. Quantitative analytical
results may also be presented and discussed where Risk Analysis
has been undertaken. Key risk themes will be identified and
discussed, as well as a trend analysis addressing changes in risk
exposure. Similar to the Summary Risk Report, the more detailed report will also end with conclusions (what does it mean)
and recommendations (what should be done), though these are
likely to be more detailed in the Detailed Risk Report. Supporting information will also be presented in appendices,
including the Risk Register, other detailed assessment results,
and supporting data.

One of the key issues in the communication plan is to identify when
stakeholders require information to be communicated, and timeliness is as
important as content. There is no advantage in getting the right information too late to use it. It is therefore important to consider when risk reporting should be undertaken in the risk process. Risk management
methodologies tend to include risk reporting as part of the final management stage, as has been done here in the Risk Monitoring, Control, and
Review phase. In reality, however, the information becomes available
incrementally throughout the risk process. For example, it is possible to list
the risks immediately after Risk Identification, and to start compiling the
Risk Register using output from the qualitative Risk Assessment. A
preliminary risk report could therefore be produced early in the risk
process, without waiting for all the details, and then reissued when more
information becomes available.

Given the need to inform project stakeholders promptly about risks
facing the project, it is recommended that risk reports should be generated

as soon as possible, even if some of the information is preliminary and will require updating later. Early reporting maximizes the time available for consideration of options, development of effective responses, and implementation of planned actions. One purpose of the risk process is to generate management space, allowing a proactive approach to dealing with uncertainty. Any delay in communicating risk information reduces the available time in which it is possible to be proactive.

A MOVING TARGET

Having communicated risk information to project stakeholders at the right level of detail and in a timely manner, agreed actions should be implemented. Without action the risk process is powerless to change the level of risk exposure. This is captured in the phrase "Use it or lose it": if the results of the risk process are not used, then any possible benefit will be lost. While it is obvious that planned responses need to be put into practice, it is important to remember that action plans include a timing element. For many risks it is possible to identify an "impact window" when the risk might occur. It is clearly essential that any proactive responses must be implemented before the start of the impact window, since there is also an "action window" in which to respond.

However, in addition to the time-stamped nature of many risk responses, it is also the case that risk information itself has a time dimension. The results of a Risk Identification exercise represent the uncertainty faced by the project at the particular point in time when the identification was undertaken. Those risks were assessed from a current perspective, formed by the prevailing circumstances at the moment of assessment. This leads to a set of preferred responses that reflect the best option at the time, given what is known about each risk, what resources are available to tackle it, the status of other project tasks, etc. Nearly all characteristics of a risk are time-dependent, including whether or not the uncertainty itself exists, through probability of occurrence and severity of impacts, to possible responses.

As time passes, so risk changes on a project, in a number of ways:

- Some identified threats may no longer exist after a period of time, either because they have been successfully avoided, or because their impact window has passed without the uncertainty actually occurring. Or perhaps worse, the uncertainty might have been removed because the threat has occurred and become an active problem or issue for the project.

- Similarly, some previously recognized opportunities may not still be there later in the project, either because they have timed-out and cannot now occur, or because they have been successfully exploited and turned into real benefits.
- Risks can get better or worse with time, in the sense that probability and/or impacts might change. This is true in terms of perception, since those responsible for assessing a risk may change their opinion of its probability and impact as they gain more information or experience on the project. It is also true in reality, since changes in external circumstances or other project variables might make a risk more or less likely or could affect the severity of its impact. These changes can occur in either direction, making a threat more or less of a potential problem, and making an opportunity more or less attractive.
- New risks can emerge that were not previously identified. This may be as a result of shortcomings in the risk identification process that failed to see some of the risks earlier, or it may be because progress on the project and changes in its environment have created new uncertainties that did not previously exist. In some cases risks may already be present inherently in a project, but may not be visible from the current perspective, emerging only after earlier project activities have been completed.

This time-based nature of risk means that risk reporting needs to minimize delay between generation and communication of risk information. Risk reports are like perishable foodstuffs which carry a "Best before" date. They are at their best when most fresh, and their quality deteriorates with time, until they go over the threshold and should not be used. Indeed, eating food after the recommended date can be positively harmful. In the same way, risk reports carry a time stamp when their contents are valid. But as time passes and things change, risk information becomes more out of date and less useful, until a time comes when it is invalid. Using the contents of a risk report at this point can be harmful, since conclusions and recommendations may be misleading. For example, the risks identified as most significant may no longer be the top risks, or preferred response strategies may no longer be the most appropriate. It is therefore important to use risk information while it is fresh. A risk report should not be put on the shelf to be read later, but should be acted on immediately if the maximum benefit is to be gained. Waiting too long before reading the report could result in taking action that is either ineffective or inappropriate.

REGULAR REVIEWS AND UPDATES

The fact that risk changes with time implies that risk reporting needs to be prompt, minimizing any delay between generating information and passing it to project stakeholders for them to act on. There is, however, another implication for the risk management process. The changing nature of risk means that a single risk assessment cannot remain valid for the lifetime of a project. Risk management must not be a one-off undertaking, but must be repeated during the project to ensure that the current risk exposure is understood and that risk management actions continue to be appropriate and effective. This means that some form of risk review should be performed as a routine part of the project management process.

Whether risks are reviewed in a separate "risk review meeting" or as part of the normal ongoing project management process depends on the level of risk management applied to the particular project, as defined in the Risk Management Plan. There is merit in holding a separate meeting focused entirely on risk, since this reinforces the importance of risk management, imposes some discipline and rigor into the risk process, and gives the project team a specific place to think and talk about how to manage the threats and opportunities associated with their project. However, the alternative of including review of risk as an agenda item for the regular project progress meeting emphasizes the integration of risk management as an essential part of the way the project is being managed, and allows other project decisions to be made in the light of current risk exposure.

Wherever they are done, risk reviews should be undertaken on a regular basis during the project. Again, the definition of "regular" depends on the level of risk process implemented on the project.

- As a minimum, risks should be reviewed at *key points in the project*, for example at project gateways, major decision points, or transitions between project phases. This ensures that the risks associated with the next phase of the project are identified and understood before proceeding with the project, including those risks carried forward unresolved from the previous phase(s).

- Alternatively, risk reviews might be undertaken at *regular intervals through the project*, for example monthly or quarterly. The frequency should reflect the perceived level of risk on the project, and may be varied during different project phases.

The problem with regular cycles of risk reviews is the tendency to assume that risks only change infrequently. This is clearly untrue, since a new risk might arise or the status of an existing risk could change at any time. It is therefore important for risk management to be seen as a continuous process, not bound by a regular review cycle. For example, the risk process should be flexible enough to allow risks to be identified whenever they are first perceived, rather than waiting for the next risk review meeting. Nevertheless regular review points are a valuable way of imposing a necessary discipline and structure on the risk process, for example ensuring that agreed actions are implemented.

A risk review should consider a number of key questions about the risks facing the project, whether it is conducted as a separate meeting or as part of a wider project meeting. It is important to:

- Address the status of existing risks (see below).
- Consider the effectiveness of planned responses and whether they are achieving the intended effect on risks.
- Assess whether any secondary risks have arisen as a result of chosen responses.
- Identify whether new responses are required to address existing risks.
- Confirm ownership of responses, recognizing that the person originally given responsibility for managing a risk may no longer be best placed to deal with it.
- Identify whether new risks have arisen since the last risk review.
- Check the effectiveness of the risk process, including whether different techniques are required to refresh or improve the process, and whether a different level of process might be appropriate.

One major purpose of the risk review is to consider the current status of each identified risk. This may be aided by defining a life cycle for a typical risk, as illustrated in Figure 6, which starts with a risk at "Draft" status. Not everything identified during a Risk Identification exercise is in fact a risk. It is therefore necessary to apply a filter to the output of Risk Identification, and some putative risks may be "Rejected," i.e., not considered to be a true risk. Items that pass the validation filter may be

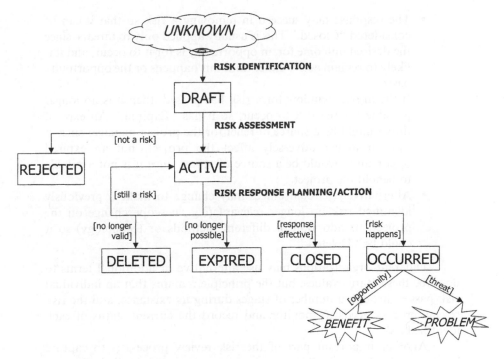

Figure 6   Life cycle of risk status values.

regarded as "Active," i.e., current uncertainties that are able to affect the project. These risks are assessed and responses are developed to address them in the most appropriate way. Following the implementation of risk responses, a number of outcomes are possible:

*   The risk may remain "Active," requiring continued monitoring and ongoing management. This means that an active threat still has the potential to affect the project adversely, whereas an active opportunity still represents a source of potential benefits.
*   If a risk happens it can take the status of "Occurred." This has a different meaning for an opportunity or a threat, since when an opportunity actually happens the benefits become available to be realized, whereas occurrence of a threat leads to existence of a problem.

- The response may succeed in removing a risk so that it can be considered "Closed." This is usually applied only to threats, since the desired outcome for an opportunity is for it to occur, and it is likely to remain active either until this happens or the opportunity passes.
- If the impact window for a risk has passed, then it is no longer possible for the risk to occur, so it has "Expired." An expired threat might be a source of relief for the project manager since it can no longer adversely affect the project, but an expired opportunity would be a source of regret since it is not available to benefit the project.
- Alternatively, circumstances may change to make a previously identified risk no longer relevant (e.g., a scope change on the project, or adoption of different methods or technology) so it could be "Deleted."

Different organizations may use alternative or additional terms to describe these status values, but the principle remains that an individual risk passes through a number of stages during its existence, and the risk review process needs to monitor and record the current status of each identified risk.

Another important part of the risk review process is to capture lessons to be learned. (These are often erroneously called "lessons learned," but the lesson is not actually learned until it is put into practice on the next project.) Some organizations may undertake this as part of project closure during the post-project review process, but there are benefits in identifying lessons as they arise, while incidents are recent and memories are fresh. In terms of the risk process, the following lessons can be identified to feed forward to future projects, or even to future phases of the same project:

> What types of risk can be identified on this type of project? Are there any generic risks that affect all similar projects?
> Which identified risks actually occurred, and why?
> Which identified risks did not occur, and why?
> Which problems occurred that could have been foreseen as threats? What preventive actions could have been taken to minimize or avoid the threat?
> Were there any missed opportunities that could have been foreseen? What proactive actions could have been taken to maximize or exploit the opportunity?

Which responses were effective in managing risks, and which were
ineffective?
How much effort was spent on the risk process, both to execute the
process and to implement responses?
Can any specific benefits be attributed to the risk process, e.g.,
reduced project duration or cost, increased business benefits, or
client satisfaction, etc?

The results from this type of lessons-learned exercise can be used to
update risk identification tools such as checklists, to incorporate preventive risk response strategies into future projects, and to improve the
effectiveness of risk management. It might also be possible to estimate
Return On Investment (ROI) for the risk process, by comparing specifically attributable benefits with process costs.

It is in the nature of risk to be dynamic, since the sources of uncertainty are constantly changing. The risk management process needs to
take this into account so that the organization can be alert and responsive
to the changing challenge of risk. Having performed a single pass through
the risk process, the organization or project team cannot rest on their
accomplishments. In the world of risk, standing still is going backward.
Risk management is not a one-off exercise, but must be repeated to
maintain a current perspective on the risk exposure. To be fully effective,
risk management must be iterative, cyclical, proactive, and forward-
looking. It should constantly be asking (and answering) questions such as:

Where are we now?
Which way are we heading?
Are we still on course?
What could affect achievement of objectives?
What can be done about it?
How does the planned response change things?

However, even the iterative nature of the risk process must be
flexible, rather than imposing a rigid five or six phases to be followed in
turn for all projects. Figure 7 indicates the cyclical nature of the risk
management process, illustrating that multiple and complex feedback
loops are required between the various steps in the process, rather than a
single pass from beginning to end. This calls for a responsive approach to
implementation of risk management on a given project, being prepared to
revisit the various stages in the light of the current level of knowledge and
understanding about the risk exposure.

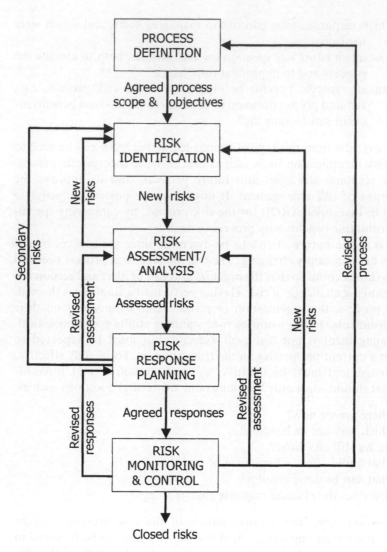

Figure 7   Cyclical risk management life cycle.

The complexity of Figure 7 should be compared with the simple cycle summary in Figure 2 of Chapter 2. While the core process should proceed in an orderly and planned fashion through the basic cycle, the need to consider the various feedbacks must not be overlooked if risk management is to be responsive and flexible enough to reflect and address changes in risk exposure.

PROBLEM AND BENEFITS MANAGEMENT

One final element is required if risk management is to achieve its aims of being fully integrated into the overall management of the project. Threats are uncertainties that if they occurred would have an adverse effect on the project; in other words, they are potential problems. Similarly, opportunities are potential benefits, uncertainties that could have a positive effect on the project. This means that the risk process should have a natural interface with two other project processes, namely, *problem management* and *benefits management*. The aim of these is to deal with risks that actually occur, including both those with negative effects (impacting threats) and those with positive effects (realised opportunities). The relationship between these processes is shown in Figure 8, which also illustrates the possible progression from problem management to crisis management and disaster recovery. Project management should encompass risk management, problem management, and benefits management, but ideally management of crises and disasters should not be required if risk management is fully effective.

While neither of the two additional processes of problem management and benefits management is strictly within the compass of risk management, it is essential that the organization wishing to manage its risks effectively also pays appropriate attention to both problems and benefits. The risk process aims to ensure that threats do not materialize into problems, but it cannot deliver with 100% efficiency, and some threats will come to pass. Without a proper problem management process the project will be vulnerable to the threats that are not addressed by the risk process. Similarly, the risk process that aims to tackle opportunities proactively will succeed in at least some cases, and an established benefits management approach is required to exploit the upside created by the risk process when opportunities become reality; otherwise the potential gain will be lost.

Many organizations lose the advantages of effective risk management by not having processes in place to deal with actual problems or unexpected

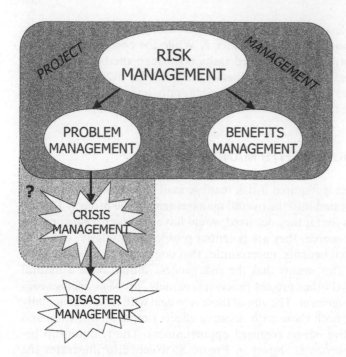

Figure 8 Risk management process relationships.

benefits. It is a waste of time if the proactive risk management process is followed by an approach to problems and benefits that is reactive, unstructured, and ineffective. This is particularly true of the upside, since many project managers seem well prepared for problem management (and even for crisis management), perhaps because they have considerable experience of operating in this mode. They seem, however, unsure how to react when faced with an unexpected opportunity, being caught by surprise and unable to take the advantage it offers. This leaves them in the "roundabouts but no swings" position discussed in Chapter 6, exposed to the downside that occurring threats bring, but without the corresponding upside of realized opportunities that could compensate for the effect of threats.

FROM ANALYSIS TO ACTION

The purpose of the risk process is to explore the uncertainty surrounding a project (or any other undertaking), understand which areas have the potential to affect achievement of objectives either positively or negatively, prioritize them for attention, assess the potential effect on the project, and develop appropriate responses. This analysis must not, however, be seen as the end result of the risk process. Risk management is an action-based process, and must lead to doing something differently as a result of the new understanding of risk delivered by the process. Decisions need to be made, strategies modified, expectations adjusted, responses designed and implemented. Without action the risk process is a waste of time.

Yet it seems that this is precisely where many organizations fail to gain the benefits promised by risk management. After time and effort have been spent in Risk Identification, Risk Assessment, and Risk Response Planning, failure to act leaves risk exposure largely unchanged, and the chances of success remain as they were before the risk process was undertaken. It is therefore essential to proceed from analysis to action, implementing agreed responses with full commitment, and paying attention to subsequent changes in risk exposure. Risk Monitoring, Control, and Review is the essential final step in the risk process, where the promised benefits are actually reaped, threats are minimized or avoided, and opportunities are enhanced or exploited. To stop before taking this final step is to waste the preceding journey.

There is no doubt that every project faces a range of uncertainties, many of which represent opportunities to improve project performance, enhance the business case, increase productivity, deliver extra capability, realize more benefits, build client confidence, complete ahead of time or under budget. The simple modifications to the typical risk process as described in the preceding chapters allow these inherent opportunities to be captured and maximized proactively as the project proceeds, rather than trusting to luck.

The project manager who is committed to project success will want to take advantage of anything that might make that more possible. Opportunity management should be treated with the same degree of serious attention as threat management, and the standard risk management process offers a framework within which both can be handled effectively together alongside each other. There is no need for a separate opportunity management process, when the principles and practice of risk management can be applied so easily to upside uncertainty as well as downside. If risk is defined as "uncertainty that can affect project objectives" and if it is accepted that this includes opportunities with upside impacts as well as threats with downside impacts, then it is a logical progression to use the same risk management process to deal with both types of risk. Simple process extensions make this possible in a way that is elegant, efficient, and effective.

REFERENCES

Bohn, R. (2000). Stop fighting fires. *Harvard Business Rev.* 78(4):83–91.

British Standard BS6079–1:2002. (2002). *Project Management—Part 1: Guide to Project Management.* London: British Standards Institute.

British Standard BS6079–3:2000. (2000). *Project Management—Part 3: Guide to the Management of Business-related Project Risk.* London: British Standards Institute.

Patterson, F. D., Neailey, K. (2002). A risk register database system to aid the management of project risk. *Int. J. Project Management* 20(5):365–374.

Williams, T. M. (1994). Using the risk register to integrate risk management in project definition. *Int. J. Project Management* 12:17–22.

III
CONCLUDING CONSIDERATIONS

9
Implementation Issues

The first part of this book laid the foundation for inclusion of opportunity within the definition of risk, and the consequential need to ensure that the risk management process also addresses opportunity alongside threat (Chapters 1–2). In the second part (Chapters 3–8) the typical risk management process was examined to determine how it might be expanded to include both types of risk, whether the potential impact of uncertainty was harmful (threat) or helpful (opportunity). Simple process enhancements have been described that allow a broadened approach to be adopted with minimal additional effort, building on the familiar techniques in use by most risk practitioners and project teams.

Although the emphasis so far has been on process, there is no doubt that this is not the whole story when it comes to implementing effective risk management. Whether the risk process includes opportunity or not, a number of factors are required in addition to the process if it is to deliver the expected benefits. This is the focus of this final part (Chapters 9–10), where key implementation issues are addressed and future developments are explored.

The role of risk management in managing projects is well recognized and understood. All projects are risky, and this requires effective management if objectives are to be achieved. Risk management is recognized as a key contributor to project success, and there is increasing interest in making it work in practice. All structured project management methodologies include risk management as one of the core areas to be addressed, and there are many standards and guidelines offering "best practice" risk processes (see Table 1 of Chapter 2).

Despite this sustained interest, most organizations and project managers would agree that risk management is not producing the expected and promised benefits. Many of those implementing some form of risk management in an attempt to address the uncertainty associated with business and projects still find their projects failing to achieve their objectives— instead they are late, over budget, or underperforming. It seems that although use of risk management is widespread, it is not always working in practice. So is it all just hype, or are there reasons for this shortfall?

Organizations wishing to implement risk management effectively will probably start by considering the three T's: techniques, tools, and training. While these are undeniably a part of the requirement, they are not enough—they are necessary but not sufficient. Of course any approach to managing risk will involve using a range of techniques, and many of these require tools to support them. The specialized nature of risk techniques and tools is likely to raise the need for training so that staff can use them properly. But these three elements alone will not make risk management effective, as amply shown by the common experience of many organizations that thought they would. In addition to these three, a number of Critical Success Factors (CSFs) must be present.

While these CSFs apply generally to any implementation of risk management, they are particularly important if the risk process is to be extended to include opportunities. Since this extension of risk management is likely to be new to many organizations, there is a real danger that they might rely too heavily on process alone, without paying adequate attention to the other factors required for effectiveness. This chapter first considers general CSFs, then looks more closely at specific issues required to support effective management of opportunities.

CRITICAL SUCCESS FACTORS FOR EFFECTIVE RISK MANAGEMENT

With widespread agreement on the techniques associated with the risk process, an abundance of supporting tools, and multiple training courses available, why is risk management apparently failing to deliver? Clearly something else is required in addition to techniques, tools, and training, to make risk management work in practice. This section discusses four CSFs that are essential for effective risk management, whether it includes opportunities alongside threats or only considers downside risks. The four CSFs are:

CSF1: clear widely accepted definitions
CSF2: a simple scalable process
CSF3: appropriate infrastructure to support the risk process
CSF4: attention to risk attitudes

CSF1: Knowing What We Mean

Everyone knows what is meant by the word "risk." Or do they? People who are not risk management specialists probably have a general concept of risk as "a bad thing" (or at least as "a potential bad thing"), but might find it hard if asked to state a clear and unambiguous definition of risk that everyone else agrees with. In fact, even risk practitioners fail to agree on what they mean by "risk," with some quite fundamental differences of opinion (as explored in the Appendix).

This is driven by the lack of consensus among published standards and guidelines. Unfortunately no single accepted definition of risk is agreed to by all, as discussed in Chapter 2 (see Table 1 of Chapter 2). All have differing definitions of risk, some of which are directly contradictory. The statement is certainly true for risk management: "The great thing about standards is that there are so many to choose from!"

Why does this matter? Because without a clear definition of risk, participants in the risk process will not be working together toward a common goal. Indeed, if their views about risk differ fundamentally, they may even be pulling in opposite directions. An agreed definition of risk is not a theoretical detail or an optional luxury. It is an essential prerequisite for a focused risk process where all interested parties share a common purpose. Before risk can be managed effectively, stakeholders and participants in the risk process need to know what "risk" means.

So what are the ingredients of risk? Is it possible to build a definition that might attract a wide consensus? This has already been discussed in Chapter 1, but is summarized briefly here. The starting point is uncontroversial: risk is related to uncertainty. A risk is not the same as a problem or issue, because it may or may not happen. This uncertainty dimension of risk is usually described using the term "probability" (though others may prefer frequency, likelihood, or chance).

But risk is not the same as uncertainty, although it is related to it. For any given project there will be uncertainties that are irrelevant and so do not pose any risk to the project. For example, "It may rain tomorrow" is not usually relevant to a project conducted entirely indoors, or "Exchange rates may fall" cannot affect projects run in a single currency. To be a risk to a particular project, an uncertainty must be able to affect the project if it happens, measured in terms of the project objectives. A risk is an uncertainty that matters. This second dimension of risk is often called "impact" (or effect, consequence, etc.), describing what could happen to the project if the risk occurs.

This gives a two-part definition of risk: "An uncertainty that, if it happens, can affect one or more project objectives."

One last question remains to be clarified—what type of effects might a risk have? This is the point where many risk standards and practitioners differ, with some saying that risk is wholly negative, others adopting a broader approach that allows risk to include both upside and downside effects, and a third group remaining neutral or agnostic on the issue (see Chapter 2).

While this debate continues, there will be no single agreed definition of risk that all accept. However, it is still vital for each organization to decide how risk will be defined for its projects. Is the focus of the risk process entirely on threats? Or does the organization expect risk management to deal equally with both threats and opportunities? The conclusion of the earlier part of this book is that "risk" properly includes both upside opportunity and downside threat, and that the risk management process should deal equitably with uncertainties that might cause harm and those that might help.

Lack of an agreed definition among those implementing risk management within an organization will produce confusion and inefficiency, with unclear scope for the risk process, diffused or wasted effort, and possible failure to identify and manage some important risks. Clear definition of what is meant by the term "risk" is a CSF for effective risk management, leading to common objectives, shared vision, and focused attention on the key risks. If the stated intention of the organization or project is to use a common risk process to address both threats and opportunities, all stakeholders and participants in the process must share this aim and be committed to implementing it in practice.

CSF2: How to Do It in Practice

In seeking to implement risk management, many organizations seem tempted to introduce complex processes with multiple tools and techniques. This often leads project managers and their teams to view risk management as a bureaucratic overhead, imposed by management to control the project, adding another layer of unnecessary analysis and reporting, but with little additional benefit. To facilitate effective risk management, the risk process must be simple enough to meet the needs of every project, and scalable to address different project types, while covering the essential steps necessary to enable risk to be managed proactively and effectively. And if risk management is to be successfully extended to include opportunities, it is even more important that the underlying process is as simple as possible, to avoid charges of imposing an additional administrative overhead burden. The typical risk process includes a number of standard steps, as outlined in Chapter 2, Figure 2, detailed in Chapters 3–8, and summarized here.

Early risk management processes started with Risk Identification as the first phase, reasoning that risks could not be managed if they were not first identified. But before risks can be identified it is important to clearly define what is meant by the term "risk," to avoid confusion and wasted effort (see CSF1 above). Defining risk as "an uncertainty that, if it happens, can affect one or more project objectives" brings together the two dimensions of uncertainty and effect on objectives. But this definition reveals a prerequisite that must be addressed before risks can be identified. If a risk can exist only in relation to project objectives, then those objectives must themselves be defined before any risks can be identified.

As a result, best-practice risk processes now start with a *Definition* phase, to be completed before Risk Identification. This ensures that project objectives are agreed, understood, and communicated within the project stakeholder community. Another purpose fulfilled by the initial Definition phase is to decide the level of detail required for the risk process, driven by the riskiness and strategic importance of the project. Some projects may only require a simple risk process, whereas others will need more in-depth risk management; the Definition phase should describe the level of detail of the process to be implemented (see CSF3 below), and document this in a

Risk Management Plan. For effective opportunity management it is important that the Definition phase unambiguously states the requirement to include opportunities within the risk process alongside threats.

Having defined project objectives and process level, the risk process can then proceed to *Risk Identification*, selecting as appropriate from a range of available techniques, including brainstorms, workshops, checklists, prompt lists, interviews, questionnaires, etc. During the Risk Identification phase, care is required to distinguish between risks and related nonrisks (e.g., problems, issues, causes, effects, etc.), and to focus attention equally on both threats and opportunities.

After risks have been identified, the next step in the risk process is qualitative *Risk Assessment*, describing characteristics of each risk in sufficient detail to allow them to be understood, allowing risks to be prioritized for further attention and action, and recording this in a Risk Register. Grouping risks using a Risk Breakdown Structure (RBS) or risk categories can also reveal patterns of risk exposure and common sources of risk. This may be followed by quantitative *Risk Analysis*, using mathematical models to simulate the combined effect of risks on overall project outcomes. The main purpose of both qualitative Risk Assessment and quantitative Risk Analysis is the same: to reveal key areas of risk exposure requiring priority treatment. The worst threats and best opportunities can then be targeted for focused management attention and action.

Following definition, identification, and assessment, the risk process should proceed to *Risk Response Planning*. In this phase, strategies and actions are determined to deal with risks in a way that is *appropriate*, *achievable*, and *affordable*. Each action should be *agreed* with project stakeholders, and *allocated* to an owner, then its effectiveness should be *assessed*. The aim is to minimize threats and maximize opportunities, to reach an overall level of risk exposure that is *acceptable*.

Planning must lead to action, so the next step involves implementation of planned actions, monitoring their effectiveness, and reporting the results to project stakeholders. This *Risk Monitoring, Control, and Update* phase is where the benefits of previous identification, assessment, and response planning are achieved, and risk exposure is actually modified on the project as a result of taking suitable action. Threats to the project are reduced or removed leading to fewer and smaller problems. Similarly, opportunities are enhanced or captured leading to increased benefits and a more successful project. This phase also includes review and update. Risk is always changing on a project as progress is made, decisions are taken, and work is completed. It is therefore important for the risk process to be iterative, regularly reviewing risk exposure, identifying and assessing new risks, and ensuring appropriate responses. The frequency of such reviews will depend on the nature of the project, but risk management should not be a one-step process, since risk itself is ever-changing.

The risk process described above is not "rocket science," but it represents simple structured common sense. One important CSF for effective risk management is to ensure that the risk process is as simple as possible while still meeting the needs of the project. The steps outlined here should form a sound basis for any risk process. Without a simple scalable process, risk management will be rejected by busy project managers as too difficult or onerous; following these simple steps will help to ensure that risk management is implemented effectively and fulfills its aims of minimizing threats, maximizing opportunities, and helping to achieve project objectives.

CSF3: The Right Level of Support

To be effective, risk management requires a structured process. However, although there is a core process to be followed (see CSF2), the level of detail required can vary from one project to another. Low-risk projects may only need a simple risk process (perhaps following the simple cycle in Figure 2 of Chapter 2), whereas more challenging projects might require a more in-depth approach (maybe implementing the level of complexity implied in Fig. 7 of Chapter 8).

In the same way, different organizations may choose to implement risk management in varying levels of detail, depending on the type of risk challenge they face. The decision over implementation level may also be driven by organizational risk appetite, and by the availability of funds, resources, and expertise to invest in risk management. The objective is for each organization to determine a level of risk management implementation that is appropriate, acceptable, and affordable. Having chosen this level, the organization then needs to provide the necessary infrastructure to support it.

Each organization needs to recognize that risk management is not a "one size fits all" discipline. While all projects are risky, and risk management is an essential feature of effective project management, there are different ways of putting risk management into practice. The organization wishing to adopt a structured approach to managing risks should therefore identify where in the range of possible implementation levels they want or need to be, recognizing that it may be necessary to offer a range of risk process levels to suit different project types within the organization.

At the most simple level is an approach known as "Slippers and Pipe" risk management. This phrase represents an informal risk process in which all the phases are undertaken, but with a very light touch. The image is of the project manager relaxing at home at the end of a busy day, sitting in front of the fire in his slippers, smoking his pipe, and reflecting on the status of his project and the challenges that he faces. (Perhaps this caricature may seem unfamiliar to the typical project manager, whose experience may be more "Stress and Prozac" than "Slippers and Pipe"!) In this informal setting, all of the phases in the risk process might be implemented as a set of simple questions. For example:

What am I trying to achieve? (Definition)
What could hinder or help me? (Risk Identification)
Which of these are most important? (Risk Assessment)
What can I do about it? (Risk Response Planning)

If these questions are followed by action and then repeated regularly (Risk Monitoring, Control, and Update), then the full risk process will have been followed, though without use of formal tools and techniques.

At the other extreme from "Slippers and Pipe" is a fully detailed risk process, which some call "Full Monty" risk management (once referring to use of Monte Carlo, but now named after the popular film!). This name is intended to represent "going all the way," with a risk process that uses a range of tools and techniques to support the various phases. For example, using this in-depth approach, stakeholder workshops leading to issue of an agreed Risk Management Plan might be used for the Definition phase, followed by multiple Risk Identification techniques involving a full range of project stakeholders singly and in groups. Use would be made of both qualitative Risk Assessment (with a Risk Register and various structural analyses) and quantitative Risk Analysis (using Monte Carlo simulation, decision trees, or other statistical methods). Detailed Risk Response Planning at both strategic and tactical levels might include calculation of risk effectiveness, as well as consideration of secondary risks arising from response implementation.

Both "Slippers and Pipe" and "Full Monty" approaches represent extremes, and the typical organization will wish to implement a level of risk management somewhere between these two. They do, however, illustrate how it is possible to retain a common risk methodology following the standard phases, while selecting very different levels of implementation.

Each organization wanting to adopt risk management consistently needs first to decide what level of process implementation is appropriate for it.

Having selected the level of implementation, it is then possible to provide the required level of infrastructure to support the risk process. This might include choosing *techniques*, buying or developing software *tools*, allocating *resources*, providing *training* in both knowledge and skills, developing *procedures* that integrate with other business and project processes, producing *templates* for various elements of the risk process, and considering the need for *support* from external specialists. The decision on the required level for each of these factors will be different depending on the chosen implementation level.

Failure to provide an appropriate level of infrastructure can cripple risk management in an organization. Too little support makes it difficult to implement the risk process efficiently, while too much infrastructure adds to the cost overhead. Getting the support infrastructure right is therefore a CSF for effective risk management, enabling the chosen level of risk process to deliver the expected benefits to the organization and its projects.

CSF4: Knowing How We Think

Risk management on projects is not performed by robots or machines. Although some university research departments are working toward this, there is currently no automated or computerized risk management expert system that can take over from people the responsibility for the risk process. It is not possible today to buy a shrink-wrapped "Risk Consultant in a Box." It is people who do all the essential steps in the risk management process, including:

> Setting risk acceptability thresholds and risk appetite
> Identifying threats and opportunities that might affect project objectives
> Prioritizing important risks by assessing probability and impacts
> Proposing appropriate responses to deal with risks
> Implementing agreed actions

All of these essential parts of the risk process require human input and judgment. Each of them is also affected by preconceptions, heuristics, and unconscious bias (discussed in more detail later in this chapter). Every person is influenced by risk attitude, which can range from risk-averse through risk-tolerant to risk-neutral or risk-seeking. This has a profound effect on how the person feels and thinks about risk, and it also influences the judgments made during the risk process. For example, a risk-averse person is uncomfortable with uncertainty, so will tend to have a low risk threshold, identify many risks, assess threats as severe and opportunities as insignificant, and prefer aggressive responses that he will implement conscientiously. A risk-seeking colleague, however, faced with exactly the same project circumstances, is likely to be more relaxed about risk, resulting in a higher risk threshold, fewer identified risks, lower assessments of threat severity, higher perceptions of possible opportunity, and more accepting responses.

This natural variation in perception may lead some to feel that risk management is all subjective and cannot be done with any degree of confidence in the results. There is, however, another approach. Like any other project variable, it is possible to identify and manage risk attitudes across the project stakeholder community. The project manager should seek to understand the risk attitudes of key stakeholders, and be aware of how this affects their participation in the risk process. Awareness leads to management, as the project manager uses the soft project management skills to influence the team and counter the effect of preconceptions and bias.

In the same way that individuals have an attitude to risk that affects their participation in the risk process, organizations also have a "risk culture" that affects the preferred approach to dealing with uncertainty (also explored further later in this chapter). There are a range of organizational risk cultures, at one extreme leading to aversion to risk, or even hostility in some cases: "We don't have risk in our projects—we're professionals/engineers/scientists . . ." This denial results in important risks being ignored, and decisions being taken without cognizance of the associated threats and opportunities. At the other end of the scale is the risk-seeking organization, and some may even be risk-addicted. A gung-ho attitude to risk will inevitably lead to disaster when the amount of risk exposure taken on exceeds the organization's ability to manage it.

The preferred risk attitude for an organization is neither risk-averse nor risk-seeking, but "risk-mature." This produces a supportive culture in the organization, which recognizes and accepts that uncertainty is inevitable, and welcomes it as an invitation to reap the rewards associated with effective risk management. Project budgets and schedules are set with the knowledge that uncertain events can influence project progress and final outcomes, but also with a commitment to provide the necessary resources and support to enable these to be managed proactively. Project managers and their teams will be rewarded for managing risks appropriately, with the recognition that some unwelcome risks occur in even the best-managed project.

"Culture" can be defined as the total of the shared beliefs, values, and knowledge of a group of people with a common purpose. It therefore has both an individual and a corporate component. For risk management to be effective, the culture must be supportive. This means that the risk attitudes of individuals must be understood and managed, and the organization's overall approach to risk must be mature.

KEY ISSUES FOR OPPORTUNITY MANAGEMENT

The four CSFs discussed above apply generally to all risk management implementations, whether they include opportunities or not. In every case there is a requirement for clear and widely accepted definitions, a simple scalable process, appropriate supporting infrastructure, and attention to the risk attitudes of individuals and the organization.

Earlier chapters have addressed the first two of these CSFs in some detail, focusing on the changes needed to definitions and process if risk management is to be successfully extended to include opportunities. The third CSF issue of infrastructure is also not contentious, as it is simply a matter of providing the necessary support to implement the selected level of risk process. The one CSF that is most influential in determining the success or failure of risk management is the fourth, namely paying proper attention to risk attitudes, both personal and corporate. And it is therefore unsurprising that this element is also particularly significant when considering how to ensure that opportunity management is effective. The remainder of this chapter addresses this issue, exploring how individual and corporate risk attitudes can affect consideration of the upside, and suggesting strategies for proactive management of this vital area. If the preferred underlying attitudes to risk of individuals and organizations can be reliably described, and if these attitudes can be consciously and deliberately modified to produce the desired appropriate outcome, then the performance of individuals, project teams, and organizations under conditions of uncertainty will be enhanced.

UNDERSTANDING INDIVIDUAL RISK ATTITUDES

Individual behavior results from the interaction between an individual's attitude and the environment in which he finds himself, as illustrated in Figure 1. When the environment is perceived as favorable or neutral,

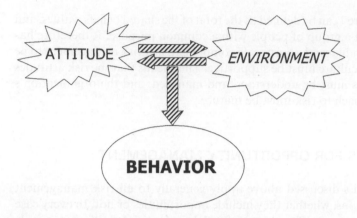

Figure 1 Attitude, environment, and behavior.

behavior is driven largely by attitude (Fig. 2). If, however, the environment is perceived as unfavorable or hostile, behavior may be forced that is
contrary to that suggested by attitude (see Fig. 3). This response is natural
and understandable. It is, however, dependent on how a particular environment in perceived by each individual. For example, an environment
that appears hostile to one may seem benign to another. Thus although it
appears at first sight that the environment is the prime determinant of
behavior, in fact it is how the environment is perceived by each person.

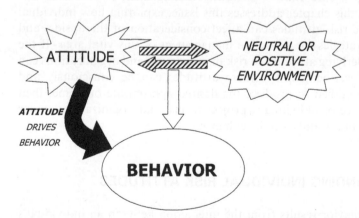

Figure 2 Behavior in neutral or positive environments.

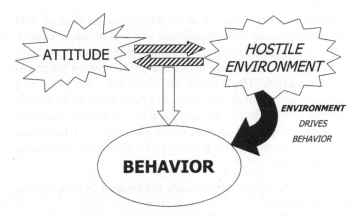

Figure 3 Behavior in hostile environments.

A further question arises about what influences behavior when the environment is uncertain. In this case the important driver of behavior is whether uncertainty is perceived as favorable, neutral, unfavorable, or hostile (Fig. 4). This perception of uncertainty is called "risk attitude."

Attitudes toward uncertainty are also affected significantly by underlying psychological influences known as *heuristics*. A heuristic in this context is defined as "an approach to inferring a solution to a problem by

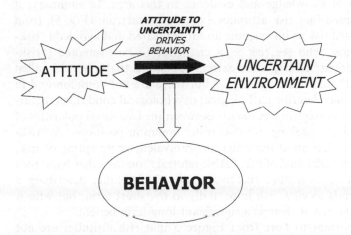

Figure 4 Behavior in uncertain environments.

reasoning from previous experience, when no relevant algorithm or data set exists." These heuristics can subconsciously and systematically introduce sources of bias when considering a situation where the answer is unknown or unfamiliar, and where a person is required to make a judgment with insufficient information. The area of risk heuristics has been well covered elsewhere by other authors, and does not need to be addressed in detail here, other than noting that if the operation of a particular heuristic is identified it can be countered and adjusted for, since all heuristics function in a systematic manner. The most typical heuristics influencing individual risk attitudes are:

"*Availability*"—more memorable events are treated as more probable or significant.

"*Representativeness*"—using similarity to stereotypes as an indicator of probability or significance.

"*Anchoring and adjustment*"—starting from an initial estimate and varying around it, even if the initial value has no objective basis in fact.

"*Confirmation trap*"—seeking and weighting evidence that substantiates a prior conviction, and ignoring contrary data.

Other heuristics operate at the organizational level and are discussed further below.

Risk attitudes have been studied by a range of academic and organizational researchers in recent years, and there is a considerable and growing body of knowledge and evidence in this area. In summary, it has been accepted that risk attitudes exist on a spectrum (Fig. 5), from those who regard risk as unwelcome and to be feared and avoided ("risk-averse") to those who see risk as a challenge to be overcome ("risk-seeking"). There are clearly more extreme positions that might be called "risk-paranoid" and "risk-addicted," but these are not common and in the worst case may describe pathological psychological conditions requiring professional corrective treatment! Between the two usual polarities of risk-averse and risk-seeking are two other common positions. A "risk-tolerant" person has an attitude that is ambivalent or accepting of risk, viewing it as a normal part of life. "Risk-neutral," on the other hand (not shown in Fig. 5), is neither risk-averse nor risk-seeking, describing a person who tends to view risk impartially in the short term, but who is prepared to take risk if there is a significant long-term benefit.

It is important to note from Figure 5 that risk attitudes are not discrete, but occupy a continuous spectrum with no clear boundaries

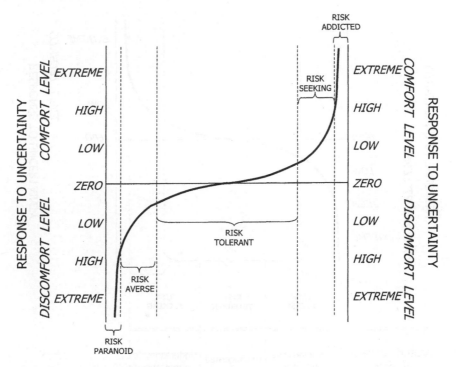

Figure 5   Spectrum of risk attitudes.

between the various headline attitudes. It is therefore possible for a particular individual to be "highly risk-averse" without being risk-paranoid, or "slightly risk-seeking" without being risk-tolerant. It is also true that the same individual may exhibit different risk attitudes under different circumstances. It is therefore a mistake to think that every person can be unambiguously labeled with a single risk attitude, although the four common terms represent real and distinct typical states. Most people have a single preferred risk attitude that represents their natural first response to uncertainty, but this can be modified by a number of factors, as discussed below.

The four basic risk attitudes are well understood and clearly defined, at least in terms of perception of risk as threat. Less work has been done on how the different risk attitudes relate to risk as opportunity. This area deserves further study and consideration, but some initial thoughts are explored below (see Fig. 6).

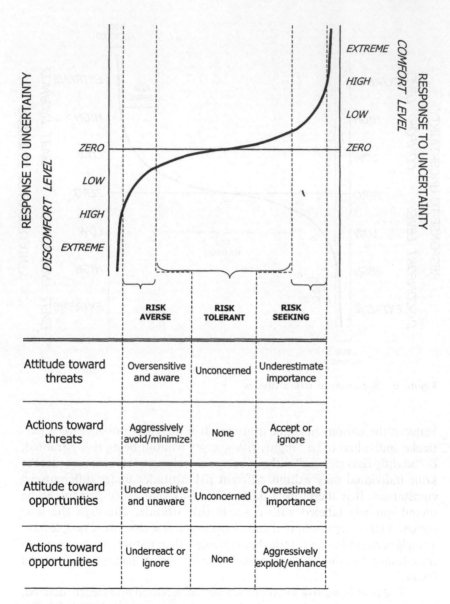

Figure 6 Effect of risk attitudes.

Risk-Averse

A *risk-averse* person feels uncomfortable with uncertainty, has a low tolerance for ambiguity, and seeks security and resolution in the face of risk. People who are risk-averse tend to be practical, accepting, and have common sense, enjoying facts more than theories, and supporting established methods of working. In terms of preferred job roles they may be comfortable as middle managers, administrators, or engineers. When applied to threats this attitude is likely to lead to increased sensitivity and overreaction, as the presence of a threat causes discomfort to people with a risk-averse attitude. This has a significant effect on all aspects of the risk process, as threats are perceived more readily by the risk-averse and are assessed as more severe, leading to a preference for aggressive risk responses to avoid or minimize as many threats as possible. When applied to opportunities, however, a risk-averse attitude is likely to lead to the opposite result, as the person may not see as many opportunities, may tend to underrate their significance, and may not be prepared to take the steps necessary to enhance or capture the opportunity. As a result, risk aversion tends to overreact to threats and underreact to opportunities.

Risk-Tolerant

Risk tolerance implies being reasonably comfortable with most uncertainty, accepting that it exists as a normal feature of everyday life, including projects and business. The risk-tolerant person tends to take uncertainty in stride, with no apparent or significant influence on his behavior. For both threats and opportunities this may lead to a failure to appreciate the importance of the potential effect of the risk on achievement of objectives, whether the impact is upside or downside, as the laissez-faire approach fails to result in proactive action. This may be the most dangerous of all the risk attitudes, since the acceptance of risk as part of the "normal situation" may mean it is not managed appropriately, leading to more problems from impacted threats, and loss of potential benefits as a result of missed opportunities. Risk tolerance may appear balanced, but progress cannot be made while remaining perfectly balanced.

Risk-Neutral

A *risk-neutral* attitude sees present risk taking as a price worth paying for future payoffs. Risk-neutral types are neither risk-averse nor risk-seeking, but rather seek strategies and tactics that have high future payoffs. They think abstractly and creatively and envisage possibilities, enjoying ideas and not being afraid of change or the unknown. As a result such people may prefer working as executives, system architects, or group leaders. For both threats and opportunities this risk-neutral approach is quite mature, focusing on the longer term and taking action only when it is likely to lead to significant benefit.

Risk-Seeking

People who are *risk-seeking* tend to be adaptable and resourceful, enjoying life and not afraid to take action. This can lead to a somewhat casual approach toward threats, as the risk seeker welcomes the challenge of tackling the uncertainty head-on, pitching his skills and abilities against the vagaries of fate. The thrill of the chase can outweigh the potential for harm, leading to unwise decisions and actions. Such people often enjoy being entrepreneurs or negotiators. During the risk process the risk-seeking person is likely to identify fewer threats as he sees these as part of normal business. Any threats that are raised are likely to be under-estimated in both probability and possible impact, and acceptance will be the preferred response. The effect of risk seeking on opportunities is quite different however. The risk seeker will be sensitive to possible opportunities, may overestimate their importance, and will wish to pursue them aggressively.

A number of instruments have been developed to assess or infer a person's preferred risk attitude, including questionnaires, 360° feedback, attitudinal surveys, behavioral interviews, job simulations, or assessment centers. Given the overlap between different risk attitudes, such assessment is necessarily an inexact science, and this area requires further study and development. Nevertheless, despite the limitations of existing assessment techniques, it is important for the individual and other project stakeholders to be aware of preferred risk attitude, so that possible biases can be identified and addressed. This also opens the possibility of modifying individual risk attitudes, once they have identified and understood.

MODIFYING INDIVIDUAL RISK ATTITUDES

It is important to realize that there is no right or wrong risk attitude. A risk-averse person should not be criticized as weak or prone to overreaction, or a risk seeker seen as reckless or lazy, since each attitude is valuable and appropriate in certain circumstances. Risk attitudes tend to be subconscious preferences developed in a person over a long period of time, partly as a result of upbringing or personal environment, and partly in response to previous experiences. Many factors can result in a particular preferred risk attitude and can influence whether a particular situation is seen as risky or not. These include:

> Level of relevant skill, knowledge, or expertise
> Perception of probability or frequency of occurrence
> Perception of impact magnitude, either severity of threat or size of opportunity
> Degree of personal control or choice in the situation
> Closeness of the risk in time or space
> Potential for personal consequences

As a result, the same individual may exhibit different risk attitudes in different situations, for example being conservative in approach to work or career (risk-averse) but undertaking free-fall paragliding as a recreational hobby (risk-seeking). And, of course, different people will respond differently to the same situation, as a result of their differing underlying risk attitude (sometimes called "perceptual dissonance")—a situation regarded as too risky by one person will be seen as acceptable by another.

Risk attitudes need to be identified, understood, and managed proactively, both by the individual and by project and line management. This requires a level of awareness that is not common in many organizations or societies. The Emotional Intelligence methodology seeks to explore awareness of attitudes within self and others, followed by conscious and thoughtful modification of these attitudes as appropriate to the requirements of a given situation. For individuals, self-reflection or use of various psychometric assessment instruments can also raise self-awareness, as can Emotional Intelligence profiling tools, such as:

> The Emotional Intelligence Individual Diagnostic Questionnaire (EIIDQ), or its team equivalent EITDQ, from the Centre for Applied Emotional Intelligence (*www.emotionalintelligence.co.uk*)
> The EQ Map from Q-Metrics (*www.qmetricseq.com*)
> The Emotional Competence Inventory (ECI) 360, developed by Richard Boyatzis and Daniel Goleman, which implements the approach from Goleman's 1998 book *Working with Emotional Intelligence*

The Simmons (SMS) EQ Profile, available via the Centre for Applied
Emotional Intelligence (*www.emotionalintelligence.co.uk*)
The BarOn Emotional Quotient Inventory (EQ-I) from the Multi-
Health Systems (MHS) Organizational Effectiveness Group
(*http://eqi.mhs.com*)

The purpose of self-awareness and recognition by others of preferred
risk attitude is not to allow individuals to be stereotyped or criticized.
Rather it is to enable a person's reaction to uncertainty to be understood
and managed appropriately, by both the person and colleagues. If behavior results from the interaction between attitude and the environment
(Fig. 1), and if risk attitude is the key driver when the environment is
uncertain (Fig. 4), then increased awareness and understanding of the
prevailing risk attitudes of participants and stakeholders can only enhance
the effectiveness of the risk management process.

It is also important to recognize that risk attitudes are not "hard-
wired" in the human brain. All attitudes are chosen situational responses,
driven by a subtle and complex set of factors, some of which are subconscious and deep-seated. But the essence of Emotional Intelligence is
that self-awareness can lead to self-regulation and management. When a
person becomes aware of his preferred risk attitude, he then understands
what his likely "gut reaction" will be when faced with an uncertain
situation. He does not, however, just have to react "from the gut," saying,
"That's just the way I am and I can't help it." An "emotionally intelligent"
person will be able to judge whether his preferred reaction is appropriate to
the situation, and if it is not, then he will be able to modify his underlying
attitude and resultant behavior accordingly. This is more than "mind over
matter" or "wishful thinking," but rather represents a mature judgment
arising from a high level of self-awareness.

In terms of risk attitudes, this is reflected in an understanding of
which attitude will produce the desired effect in a given situation, and
consciously modifying one's attitude as appropriate, even if this means
adopting an approach that is counterintuitive. For example, a naturally
risk-averse manager may realize the need to take a degree of risk that
exceeds the intuitive comfort level to gain the required business benefit, so
the manager can choose a more risk-seeking approach in that situation
even if it would not be the preferred way to proceed. The key is to respond
rather than react, thinking about the issues involved and making an
informed choice. Of course, an individual may require input and support
to diagnose and modify risk attitudes, and self-reflection may not be

enough alone. Mentoring and coaching may be useful in this context to guide people toward self-awareness and provide strategies to support attitudinal and behavioral change.

An important question here is whether it is possible for an individual to have different risk attitudes toward threats and opportunities. For example, can one person show risk aversion toward threats and simultaneously be risk-seeking in relation to opportunities? This does not appear to have been addressed by current research, and might be worth exploring in the future. Intuitively one might expect an individual to have one main preferred risk attitude, which is then applied differently in situations where there is potential for upside or downside impact, particularly for people at the two extremes of risk aversion or risk seeking. Thus a risk-averse person might tend to both overstate the significance of threats and underplay potential opportunities. A risk seeker may take a too casual approach to threats but be very keen to exploit opportunities.

This all emphasizes the importance of teamwork in the risk process, as in wider project management and business. Once individual risk attitudes are identified and understood, it becomes possible to build risk-balanced teams where the strengths of one complement the weaknesses of another. The risk-averse person can be relied on to challenge plans and strategies, looking for threats and weak spots, and testing the feasibility of proposed solutions. This plays to his strength of seeking certainty and security. The risk-seeking colleague can then be tasked with improving the proposed solution, exposing opportunities to do better-faster-cheaper. The risk-neutral team member will test both of these positions for their long-term benefit and advantage to the project and the organization, while the risk-tolerant offers a balanced view from the center. The risk-balanced team can be expected to perform more effectively than teams where a single risk attitude predominates.

Building risk-balanced teams will require development of reliable instruments for diagnosing risk attitudes, which are consistent across age, gender, industry type, nationality, etc. Work is in progress in this area, but there is currently no widely accepted framework for such risk attitude diagnosis.

One further issue relating to individual risk attitudes toward threats and opportunities is to recognize the natural human tendency to be concerned about potential hazards, particularly among those whose preferred risk attitude is risk-averse. Some view this as a natural defense mechanism arising from human evolutionary history ("survival of the threat-aware"), and others see it as arising from innate intellectual skepticism in the technocratic age ("that will never work"). Experience of conducting risk reviews with project teams in a variety of industries and countries suggests that it often seems easier for people to identify threats than opportunities, although this may be a function of the types of people working on such projects. Potential downside impact takes priority over upside in the minds of most people working on projects, perhaps driven by the focus of projects on achieving set objectives within a constrained environment.

Where it exists, the preference to focus first on threats needs to be tackled if the risk process is to be extended successfully to include both threats and opportunities. Left unchecked or unmanaged, such a preference can result in Risk Identification that concentrates almost exclusively on threats, despite encouragement or direction from facilitators or managers to include opportunities. Rather than simply shouting louder and haranguing people to remember opportunity, a positive strategy is to consciously address threats first in any Risk Identification exercise, before dealing with opportunities. Once threat-focused people are secure and confident that threats have been identified and will be managed proactively and effectively, then they will feel able to move on to consider opportunities. This suggests that effective threat identification may be a psychological prerequisite for opportunity identification for such people. It should of course be recognized that this threat focus is less likely to affect individuals with a different risk attitude, though such people are not usually the majority in most project teams.

People can also be trained to look actively for opportunities, modifying this threat-preference risk attitude in the same way as they can modify other attitudes. This is more than "the power of positive thinking," but instead arises from a mature understanding that opportunity exists and can be proactively identified and effectively managed. With practice, it can become as natural to explore possible upsides as it is to look for potential failures. The aim is to engender a frame of mind that consistently and persistently seeks the upside even in the face of adversity and threat. This approach follows the advice of British prime minister Sir Winston Churchill (1874–1965), who said, "Success is the ability to go from one failure to another with no loss of enthusiasm," echoing the proverb "Success means trying once more than the number of failures." This approach is particularly helpful to countering the risk-averse perspective that tends to overlook opportunities and focus unduly on threats, and it can result in a more healthy degree of realistic opportunity taking.

One final note of caution must be expressed when considering the need to encourage people to look for opportunities. Sometimes this search can lead to unwarranted optimism that results in unwise choices or unacceptable risk taking. The 108th Annual Convention of the American Psychological Association (APA) held in Washington, DC, in August 2000 took the theme "The (Overlooked) Virtues of Negativity." This suggested that relentless optimism can be dangerous, and instead recommended adoption of "defensive pessimism" as a strategy for coping with uncertainty. The proponents of this approach suggest that preparing for the worst can release people to do their best, as the reduction in anxiety levels allows them to perform better. It is also, of course, an effective means of identifying threats. While this may be particularly applicable to cultures prone to positivism and the "can-do" attitude, defensive pessimism is also relevant to modifying risk attitudes, as it can be used to temper the natural optimism of the risk-seeking person. As U.S. politician Robert Kennedy (1925–1968) said, "Only those who dare to fail greatly can ever achieve greatly." By allowing the worst case to be considered, defensive pessimism can release risk-seeking individuals to a higher level of achievement than their natural risk attitude might otherwise allow.

UNDERSTANDING CORPORATE RISK CULTURE

While the area of individual risk attitude has been well characterized and understood, the parallel issues relating to corporate risk culture are less well recognized. It is often not readily accepted that an organization can have a distinctly defined approach to uncertainty, or that it is possible for this to be determined and modified in a similar way as individual risk attitudes. This is partly driven by the wider discussion over whether an organization as an entity can have a "company culture" or display "organizational psychology" or "corporate behavior." A definition of culture was given earlier in this chapter as "the total of the shared beliefs, values, and knowledge of a group of people with a common purpose," with both an individual and a corporate dimension. Specifically in the arena of approaches to risk, however, it seems clear that a given organization can adopt a distinct risk attitude, whether this forms part of a broader culture or not.

At the simplest level, organizations or their component parts can be divided into the same categories as individuals, on a spectrum from risk-averse through risk-tolerant and risk-neutral to risk-seeking (see Fig. 5). One of the drivers of the risk attitude of an organization is the characteristics of its industry sector; for example, risk aversion typically may be displayed by providers of banking and financial services, nuclear and energy sectors, and government departments. Risk-seeking organizations might include venture capital companies, the pharmaceutical and biotech sector, or marketing agencies, as well as small entrepreneurs and start-ups. In the same way that individual risk attitudes influence differently the approach to threats and opportunities (as discussed above), corporate risk attitudes have the same effect. Thus an organization that is predominantly risk-averse will tend to be oversensitive to threats and will take aggressive steps to protect against them, and the same organization may also fail to take the actions necessary to capture opportunities as a direct result of their unwillingness to take risk.

As with individuals, a number of heuristics can affect corporate risk attitude. These are underlying innate paradigms deep within the organizational psyche (as distinct from deliberate or conscious choices to distort or obscure the truth), and often individual managers or decision makers are unaware of their existence or influence. The most common corporate risk heuristics relate to group dynamics that operate when making decisions under conditions of uncertainty, including:

> "*Groupthink*"—a mode of thinking that people engage in when they are deeply involved in a cohesive group, when the members' strivings for unanimity override their motivation to realistically appraise alternative courses of action (defined by psychologist Irving Janis (1918–1990) following the notorious 1959–1961 Bay of Pigs fiasco, when members of the U.S. Kennedy administration all agreed with the apparent consensus to invade Cuba while privately holding serious reservations)
> "*Risky shift*"—the tendency of a group to be more risk-seeking than its constituent individuals, driven by "safety in numbers" and the lack of individual accountability, as well as the tendency of risk-seeking group members to be more vocal and persuasive
> "*Cautious shift*"—the opposite of "risky shift," when the group becomes more conservative than its individual members, if no one is prepared to take responsibility for risk taking, or if the group tends to a middle course while seeking consensus
> "*Cultural conformity*"—making decisions that match the perceived company ethos or style (often expressed as "the way we do

things around here"), or that are compliant with accepted social or cultural norms

"The Moses factor"—when the group follows the example of a charismatic leader and adopts the leader's preferred risk attitude even when it contradicts the personal preferences of group members

Another important factor influences corporate risk culture, namely the prevailing risk culture of the society in which the organization exists. In the same way that organizations can display a coherent risk attitude, it is possible to define a preferred approach to risk within a given social setting. Pioneering work by Geert Hofstede (1928–) in the 1980s explored a number of characteristics of culture across a wide range of countries. One element examined was what Hofstede called "Uncertainty Avoidance," defined as "the extent to which members of a culture feel threatened by uncertain or unknown events," or "the degree to which people seek to avoid uncertainty or ambiguity." His research used a number of measurable factors to calculate an "Uncertainty Avoidance Index " (UAI) for different countries, as shown in Table 1. Relating this to risk attitude, some have concluded that a high UAI corresponds to risk aversion, and low UAI represents risk seeking, but more recently Hofstede has made it clear that this is an oversimplification.

Hofstede characterized high-UAI countries as having a higher anxiety level, concerned about the future, driven by fear of failure, committed to hierarchical structures, resisting change, and seeking consensus. This includes what Hofstede called the "Latin cluster," containing Italy, Venezuela, Colombia, Mexico, and Argentina. On the other hand, low-UAI countries appear to have a lower anxiety level, be prepared to take life a day at a time, driven by hope of success, prepared to bypass hierarchy where justified, prepared to embrace change, and recognizing the value of competition and conflict. The so-called "Anglo cluster" match these characteristics, including Great Britain, the United States, Canada, Ireland, Australia, New Zealand, South Africa, India, and the Philippines.

There are some criticisms of Hofstede's original work, including the fact that it was based on survey data (over 116,000 responses) from 1968 and 1970, and societies have changed dramatically since then (although it can be argued that the deep underlying national cultural characteristics change more slowly). Calculation of UAI was also based on only three diagnostic questions, so the limited empirical database may have resulted in oversimplified inferences that may perhaps go beyond what the data can

Table 1 Uncertainty Avoidance Index (UAI) by Country/Region

Country	UAI Score	Country	UAI Score
Greece	112	Equador	67
Portugal	104	Germany	65
Guatemala	101	Thailand	64
Uruguay	100	Iran	59
Belgium	94	Finland	59
Salvador	94	Switzerland	58
Japan	92	West Africa	54
Yugoslavia	88	Netherlands	53
Peru	87	East Africa	52
France	86	Australia	51
Chile	86	Norway	50
Spain	86	South Africa	49
Costa Rica	86	New Zealand	49
Panama	86	Indonesia	48
Argentina	86	Canada	48
Turkey	85	U.S.A.	46
South Korea	85	Philippines	44
Mexico	82	India	40
Israel	81	Malaysia	36
Colombia	80	Great Britain	35
Venezuela	76	Ireland	35
Brazil	76	Hong Kong	29
Italy	75	Sweden	29
Pakistan	70	Denmark	23
Austria	70	Jamaica	13
Taiwan	69	Singapore	8
Arab countries	68		

Source: Hofstede, 1982.

support. Some of Hofstede's conclusions are consistent with other work, but there has been no confirmatory study that is directly comparable in scope or scale. A further limitation is the fact that the original work was done all in one large multinational company in an attempt to focus the work entirely on national cultural differences and screen out influences of differing organizational culture. This probably skewed the data but in a way that is hard to identify and correct for.

Clearly an organization is likely to be influenced by the prevailing culture of the country in which it operates. As a result, one might expect to be able to correlate corporate risk culture with national UAI score. There are a number of limitations to this however. Foremost of these is the fact that most countries are not monocultures, and multiculturalism has increased dramatically in recent years. This means that there is unlikely to be a single culture within a given country, although a majority perspective may dominate. Second, many organizations are multinational, operating across a range of countries, so they are not subject to a single cultural influence. Although local offices of multinational corporations often display distinct cultures that reflect the host country, it is less clear whether an overarching corporate culture is driven by one or more national influences, for example the culture of the country that is home to headquarters. As for many countries, most multinational corporations are not monocultures. Indeed, even organizations operating wholly within a single country may have varying subcultures or microcultures in different departments, divisions, or locations. This raises the key point that it is much more useful to understand risk attitudes and culture at an individual and team level than at the level of the organization or nation, although the higher levels undoubtedly have an influence.

MODIFYING CORPORATE RISK CULTURE

Drivers of corporate risk culture are many and complex, arising from external influences such as host country UAI, internal pressures such as group dynamics risk heuristics, and the underlying risk attitude characteristic of the organization's industry sector. Nevertheless, in the same way that awareness of individual risk attitude opens the door to modification, so an organization that understands its preferred approach to risk and that has identified the key influences on that approach can undertake steps to modify the corporate risk culture.

The first step in achieving this transformation of organizational culture is to start with the senior leadership team of the organization, ensuring that they are aware of the issues and their implications for the business. Executive coaching might be a valuable strategy to adopt at this stage, to increase the risk awareness of individual senior managers and leaders within the business, and to create a pull from the top that encourages and models change in risk attitude.

As for individual risk attitudes, the existing corporate risk culture is not immutable. Awareness is the necessary first step toward change, and an organization can then respond to modify its corporate risk attitude to match the demands of particular situations. A strategic audit of corporate risk attitude can be undertaken to diagnose the presence and strength of various drivers of risk culture described above, defining routes to improvement and development as part of an overall change program. This might require minor adjustments to the way the business operates, or could involve a more wide-ranging organizational redesign, aiming to make the organization alert and responsive to both upside and downside uncertainty. For example, when facing a recession an organization can adopt appropriate risk-averse strategies to protect the core business while remaining alert to possible expansion or diversification opportunities that would demand the ability to take risks.

The key is for an organization to respond proactively to the challenges of a given set of circumstances so that it is in the best position to exploit upside uncertainties while minimizing the downside. And one key discriminator of this is the organization's attitude toward opportunity, and whether it is prepared to modify its risk attitude to enable it to take advantage of those risks that offer the possibility of enhanced benefits and gain. By understanding its preferred or innate risk attitude, an organization can make the cultural changes necessary to respond appropriately to its uncertain environment to minimize and avoid threats while simultaneously enhancing and capturing opportunities.

A number of existing approaches can be adopted to assist the organization in modifying its risk culture. For example, techniques designed to stimulate or support creativity and innovation are well suited to encouraging organizations to think positively, see opportunities, and develop strategies to capture benefits. One of the foremost innovation techniques is the Theory of Solving Inventive Problems (known as TRIZ from its Russian acronym). TRIZ was developed in 1946 by Russian engineer Genrich Altshuller (1926–1998) from his empirical study of the principles underlying innovation. After examining over 400,000 patents he formulated a number of laws to express common features. This has led to the discipline of "systematic inventive thinking," which holds that inventiveness and creativity can be taught. Applying such thinking is likely to enhance an organization's ability to identify and capture opportunities proactively and effectively.

The approach known as Appreciative Inquiry also aims to release creativity by building a "positive change core" within an organization. Appreciative Inquiry uses a range of techniques to expose options and achieve the best possible outcome, in both people and organizations. The central theme involves asking structured and systematic questions to apprehend, anticipate, and heighten positive potential—in other words, to identify and exploit opportunity. Having defined the potential, visualization techniques can be employed to "see" the desired future and envisage how it might be created. Such techniques include Scenario Painting, Rich Pictures, and Storytelling. This is an area currently undergoing active development, and should be monitored closely by organizations looking for ways of diagnosing and modifying their risk culture to create a climate more conducive to opportunity management.

Other similar approaches can be adopted, drawing on the growing resources from the areas of organizational development and emotional literacy. These include a range of commercial products including those listed in the previous section. Development of emotional literacy across an organization can encourage the ability to handle uncertainty positively.

Before leaving this area, it is worth addressing a comment often heard in the risk management arena, namely that "Effective risk management requires a blame-free culture." The issue of blame is of particular relevance to the risk process that aims to minimize negative impact and maximize positive results, since the occurrence of problems or loss of benefits can easily be seen as resulting from a failure of risk management. The organization's attitude to unwelcome results or performance is certainly part of the underlying risk culture, but it must be questionable whether it is either possible or desirable to be truly "blame-free." If the organization genuinely attempts to implement a no-blame culture it forfeits any personal accountability, undermines responsibility, denies management sanction or discipline, and leads to tolerance of antisocial or unprofessional behavior. The ability to allocate responsibility and accountability for wrong or inappropriate actions or decisions is fundamental to the learning organization, forming part of the lessons-learned process.

It is therefore necessary to balance a nonpunitive learning environment with the requirement to hold people accountable for their decisions and actions. This requires development of an open and fair culture, where allocation of responsibility and accountability is seen as the just dessert and a necessary corrective. It should be possible for an organization to manage its use of accountability to reach the optimum balance. The situation is similar to "stress," which is not entirely bad—some stress is necessary to function normally (so-called "eu-stress") but too much stress causes problems ("distress"). Perhaps organizations should permit a level of "eu-blame" to allow the normal functioning of management discipline and personal/professional accountability, while being careful to avoid "dis-blame."

ORGANIZATIONAL RISK MANAGEMENT MATURITY

The goal for the risk-aware organization is to be able to modify its risk attitude appropriately given the changing environment in which it operates. This requires a flexible corporate risk culture that is neither always anxiously risk-averse nor unflinchingly risk-seeking, but that can be described as "risk-mature."

Maturity in risk management capability has a number of dimensions, and considerable work has been done in recent years to develop frameworks to define these so that organizations can benchmark themselves against objective standards, understand their strengths and weaknesses, and create a structured route to improvement in risk management capability and effectiveness.

One such framework is the Risk Maturity Model (RMM), which was developed to meet the needs of organizations wishing to compare their management of risk with best practice or against competitors using an accepted benchmark for organizational risk capability. Development of the RMM drew on established concepts from existing models such as the Capability Maturity Model (CMM) from Carnegie-Mellon Software Engineering Institute and the European Foundation for Quality Management (EFQM) Business Excellence Model. Other models for assessing specific aspects of risk management capability have since been developed based on the RMM, including versions to address safety risk (SRMM) and business risk (BRM3). The RMM describes four levels of increasing risk management capability, termed *Naïve, Novice, Normalized,* and *Natural.* The aim is to provide a structured route to excellence in risk management, with recognizable stages along the way that organizations can use to benchmark themselves against. The various levels are defined as follows:

- The *Naïve* risk organisation (RMM Level 1) is unaware of the need for management of risk, and has no structured approach to dealing with uncertainty. Management processes are repetitive and reactive, with little or no attempt to learn from the past or to prepare for future threats or uncertainties.

- At RMM Level 2, the *Novice* risk organization has begun to experiment with risk management, usually through a small number of nominated individuals, but has no formal or structured generic risk processes in place. Although aware of the potential benefits of managing risk, the Novice organisation has not effectively implemented risk management and is not gaining the full benefits.
- The level to which most organizations aspire when setting targets for management of risk is captured in RMM Level 3, the *Normalized* risk organization. At this level, management of risk is built into routine business processes and risk management is implemented on most or all projects. Generic risk processes are formalized and widespread, and the benefits are understood at all levels of the organization, although they may not be fully achieved in all cases.
- Many organizations would probably be happy to remain at Level 3, but the RMM defines a further level of maturity in risk management capability as a stretch target, termed the *Natural* risk organization (Level 4). Here the organization has a risk-aware culture, with a proactive approach to risk management in all aspects of the business. Risk information is actively used to improve business processes and gain competitive advantage. Risk processes are used to manage opportunities as well as potential negative impacts.

Each RMM level is further defined in terms of four attributes, namely *culture*, *process*, *experience*, and *application*. These allow an organization to assess its current approach to risk management against agreed criteria, set realistic targets for improvement, and measure process toward enhanced risk capability. Characteristics of the four attributes at each RMM Level are listed in Table 2 and summarized as follows:

For the Level 1 Naïve organization, the attributes are all at the lowest level. The *culture* is resistant to change and the need for risk management is not recognized. There are no risk *processes*, no *experience* of using risk management, and no *application* to projects or the business.

The *culture* of the Level 2 Novice organization is not fully convinced of the benefits of risk management and tends to see it as a necessary overhead. *Processes* are rather ad hoc and their effectiveness depends on the limited *experience* of a few key

Table 2 Attributes of RMM Levels

	Level 1—Naive	Level 2—Novice	Level 3—Normalized	Level 4—Natural
Definition	Unaware of the need for management of risk No structured approach to dealing with uncertainty Repetitive and reactive management processes Little or no attempt to learn from past or to prepare for future	Experimenting with risk management, through a small number of individuals No generic structured approach in place Aware of potential benefits of managing risk, but ineffective implementation, not gaining full benefits	Management of risk built into routine business processes Risk management implemented on most or all projects Formalized generic risk processes Benefits understood at all levels of the organization, although not always consistently achieved	Risk-aware culture, with proactive approach to risk management in all aspects of the business Active use of risk information to improve business processes and gain competitive advantage Emphasis on opportunity management ("positive risk")
Culture	No risk awareness Resistant/reluctant to change Tendency to continue with existing processes	Risk process may be viewed as additional overhead with variable benefits Risk management used only on selected projects	Accepted policy for risk management Benefits recognized and expected Prepared to commit resources to reap gains	Top-down commitment to risk management, with leadership by example Proactive risk management encouraged and rewarded
Process	No Formal Process	No generic formal processes although some specific formal methods may be in use	Generic processes applied to most projects	Risk-based business processes

Process effectiveness		Process effectiveness depends heavily on the skills of the in-house risk team and availability of external support	Formal processes, incorporated into quality system Active allocation and management of risk budgets at all levels Limited need for external support	"Total Risk Management" permeating entire business Regular refreshing and updating of processes Routine risk metrics with constant feedback for improvement
Experience	No understanding of risk principles or language	Limited to individuals who may have had little or no formal training	In-house core of expertise, formally trained in basic skills Development of specific processes and tools	All staff risk-aware and using basic skills Learning from experience as part of the process Regular external training to enhance skills
Application	No structured application No dedicated resources No risk tools	Inconsistent application Variable availability of staff Ad hoc collection of tools and methods	Routine and consistent application to all projects Committed resources Integrated set of tools and methods	Second-nature, applied to all activities Risk-based reporting and decision-making State-of-the-art tools and methods

individuals who may have little formal training. Risk management *application* is inconsistent and patchy.

Level 3 organizations have Normalized risk into their way of operating, with a *culture* that recognizes the existence of risk and
expects to reap benefits from managing it. Generic and formal
processes are in place, with the necessary resources available,
and staff have adequate *experience* and expertise to undertake
effective risk management. *Application* is routine and consistent across all projects.

At Level 4 Natural, a risk-aware *culture* drives the organization into
proactive risk management, seeking to gain the full advantages
of the uncertain environment. Best-practice *processes* are implemented at all levels of the business, with regular updating and
active learning from previous projects. All staff have a degree
of *experience* of using risk processes to assist their tasks, and
application is widespread and second-nature across all areas.

Lower-level diagnostic characteristics can also be defined below each
RMM attribute (see Fig. 7), allowing a detailed assessment of strengths
and weaknesses, rather than assessing "overall risk management maturity" simplistically as a single number representing RMM Level.

One key characteristic of increasing maturity as measured by the
RMM is the approach taken to opportunities within the risk management
process. Indeed the way opportunities are treated by an organization may
be a good prima facie indicator of overall risk management maturity:

At RMM Level 1 there is no recognition of the existence of any type
of risk or the need for risk management, so opportunities are
obviously not managed proactively by the organization.

RMM Level 2 adopts an ad hoc approach to managing risk, and this
is commonly typified by a concentration on reducing or avoiding the most significant threats, particularly in the technical or
safety areas, with no visibility of or attention to opportunities.

More mature organizations at RMM Level 3 use a structured approach to risk management with formal processes routinely
applied to all projects, but again the tendency is to concentrate
on threat management, with little attempt to manage opportunities consistently.

It is only at the most mature level within the RMM (Level 4) that
opportunity management is usually found as an explicit and
integral part of the risk management process.

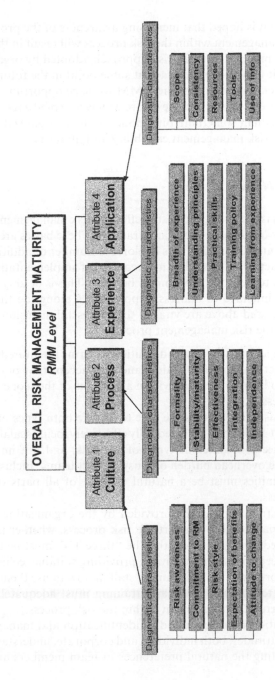

Figure 7 Risk Maturity Model attributes and diagnostic characteristics.

Of course it is hoped that increasing awareness of the proper role of opportunity management within the risk process will result in this element being included more often in the risk approach adopted by organizations. Consequently it may prove necessary at some point in the future to revise the diagnostic characteristics of the RMM to place opportunity management in RMM Level 3, as it becomes more widely accepted as common best practice, rather than something only done by those organizations at the leading edge of risk management capability and maturity.

MAKING IT WORK!

A good process is not enough to ensure effective risk management, even if it is supported by techniques, tools, and training. These basics are necessary but not sufficient. This chapter has explored a number of additional Critical Success Factors (CSFs) that are required if implementation of risk management is to deliver the promised benefits. This is doubly true if the risk process is intended to address opportunities alongside threats. The four CSFs discussed above are vital to the successful inclusion of opportunities within the risk management process:

1. Clear and unambiguous definition of terms is required so that all project stakeholders understand and accept that opportunities are to be fully included within all aspects of the scope of the risk process.
2. The risk process must be able to handle both types of risk (i.e., threat and opportunity) equally effectively, being scalable to meet the needs of different types of projects, and not imposing an undue overhead burden on busy project teams. Inclusion of opportunities must be a natural element of all parts of the risk process.
3. Infrastructure must be provided by the organization at an appropriate level to support the risk process, whether it is implemented simply or in depth. The "three T's" must be tailored to the needs of the organization, providing suitable techniques and supporting tools, and training staff in how to use them properly. Such techniques, tools, and training must adequately support opportunity management within the risk process.
4. Attention is also required to identification and management of risk attitudes, both individual and corporate, understanding and adjusting the natural preferences of team members and the or-

ganization when faced with uncertainty, including both down-side threats and upside opportunities.

Of these four CSFs, the last is the most influential and the most difficult, since it touches most closely on personal and private issues. It is, however, the most important to get right if risk management is ever to work as it should. Particularly when considering opportunities, it is necessary to understand the reaction of oneself, other stakeholders, and the organization at large, as these attitudes exert a significant influence on the approach adopted to every opportunity. Risk attitudes must be managed proactively, as with every other aspect of the risk process, and they should be modified where necessary to create the optimum environment for effective risk management, including both threats and opportunities. This includes both individual risk attitudes and the prevailing corporate risk culture, both of which can make it impossible to deal with threats and opportunities appropriately. Fortunately there are a range of strategies that can be adopted to assess and amend risk attitudes at both individual and corporate levels, so there should be no excuse for allowing unconscious underlying preferences or hidden agendas to hinder effective management of both threats and opportunities. The aim is for individuals to become more risk-aware and for organizations to become more risk-mature, modifying risk attitudes to meet the varying challenges of the changing project and business environment.

Addressing these issues will help to ensure that the risk process works in practice, operating smoothly and effectively, generating buy-in and commitment from participants and stakeholders, and delivering the promised benefits. If the organization wishes to extend the risk process as described in Part II of this book, attention to these CSFs is essential. It is not enough to rely on implementing process changes or providing techniques, tools, and training. Identifying, enhancing, and exploiting opportunities is too important to leave to chance, and should not be done half-heartedly. Instead it must be done properly if it is to be done at all, with attention to the issues that can make the difference between going through the motions and achieving the intended results.

REFERENCES

Altshuller, G. (1996). *And Suddenly the Inventor Appeared: TRIZ, the Theory of Inventive Problem Solving.* 2nd ed. Worcester, MA: Technical Innovation Center.

Altshuller, G. (1997). *40 Principles: TRIZ Keys to Technical Innovation*. Worcester, MA: Technical Innovation Center.

Artto, K. A., Hawk, D. L. (1999). Industry models of risk management and their future. In: *Proceedings of the 30th Annual Project Management Institute Seminars, Symposium*, presented in Philadelphia, USA, 11-13 October 1999.

Baldwin, D. G. (2001). How to win the blame game. *Harvard Bus. Rev.* 79(7): 55–62.

Bannister, J. (January/February 2001). Risk paranoia. In: *InfoRM, J. UK Inst. Risk Management*, 12–14.

Cabanis-Brown, J. (1999). The human task of a Project Leader: Daniel Goleman on the value of high EQ. *PM Network* 13(11):38–41.

Casper, C. M. (2002). Using Emotional Intelligence to improve project performance. In: *Proceedings of the 33rd Annual Project Management Institute Seminars, Symposium* (PMI 2002), presented in San Antonio, USA, 7-8 October 2002.

Checkland, P., Scholes, J. (1990). *Soft Systems Methodology in Action*. New York: John Wiley.

Chernis, C., Goleman, D., eds. (2001). *The Emotionally Intelligent Workplace*. San Francisco: Jossey-Bass.

Cooper, D. (1997). Evidence from safety culture that risk perception is culturally determined. *Int. J. Project Business Risk Management* 1(2):185–202.

Cooperrider, D. L., Srivastva, S. (1987). Appreciative Inquiry in organisational life. *Res. Organisational Change Development* 1:129–169.

Druskat, V. U., Wolff, S. B. (2001). Building the emotional intelligence of groups. *Harvard Business Rev.* 79(3):80–90.

Fischhoff, B. (1985). Managing risk perceptions. *Issues Sci. Technol.* 2(1): 83–96.

Fischhoff, B., Lichtenstein, S., Slovic, P., Derby, S., Keeney, L. (1981). *Acceptable risk*. Cambridge, UK: Cambridge University Press.

Fitzgerald, S. C., Murrell, K. L., Miller, M. G. (2003). Appreciative Inquiry: accentuating the positive. *Business Strategy Rev.* 14(1):5–7.

Franklin, J., ed. (2001). *The Politics of Risk Society*. Cambridge, UK: Polity Press.

Furedi, F. (2002). *The Culture of Fear: Risk Taking and the Morality of Low Expectation*. Rev. ed. New York: Continuum International Publishing Group.

Garvin, D. A., Roberts, M. A. (2001). What you don't know about making decisions. *Harvard Business Rev.* 79(8):108–116.

Gault, W. (August 2002). From blame to aim. *InfoRM J. UK Inst. of Risk Management*, 12–14.

Goleman, D. (1995). *Emotional Intelligence: Why It Can Matter More Than IQ*. London: Bloomsbury Publishing plc.

Goleman, D. (November–December 1998). What makes a leader? *Harvard Business Rev.*, 93–102.

Goleman, D. (1998). *Working with Emotional Intelligence*. London: Bloomsbury Publishing plc.

Goleman, D. (2003). *Destructive Emotions*. London.: Bloomsbury Publishing plc.

Goleman, D., Boyatzis, R., McKee, A. (December 2001). Primal leadership: the hidden driver of great performance. *Harvard Business Rev.*, 42–51.

Greenwood, M. (1998). Measuring risk behaviour: an introduction to the Risk Behavior Profiler. *Int. J. Project Business Risk Mgt.* 2(3):273–288.

Hall, D. (2002). Risk management maturity level development. Available from PMI Risk SIG website *www.risksig.com/articles/rm%20report%20 final%20version%2012.doc.*

Hall, E. (1990). *Understanding Cultural Differences.* Yarmouth, MA: Intercultural Press.

Hall, E. M. (1998). *Managing Risk—Methods for Software Systems Development.* Reading, MA: Addison Wesley Longman.

Hammond, J. S., Keeney, R. L., Raiffa, H. (September/October 1998). The hidden traps in decision making. *Harvard Business Rev.*, 47–58.

Hammond, S.Royal, C. eds. (1998). *Lessons from the Field: Applying Appreciative Inquiry.* Plano, TX: Practical Press Inc.

Hillson, D. A. (1997). Towards a risk maturity model. *Int. J. Project Business Risk Mgt.* 1(1):35–45 [The Risk Maturity Model was a concept of, and was originally developed by, HVR Consulting Services Limited in 1997. All rights in the Risk Maturity Model belong to HVR Consulting Services Limited.]

Hillson, D. A. (1998). Benchmarking organisational risk capability using the Risk Maturity Model. *Project* 10(8):13–14.

Hillson, D. A. (2000). Benchmarking risk management capability. In: *Proceedings of the 3rd European Project Management Conference*, presented in Jerusalem, Israel, 12-14 June 2000.

Hillson, D. A. (July–December 2002). Critical success factors for effective risk management. *PM Rev.*, Four-part Series.

Hillson, D. A. (April 2002). What is risk? Towards a common definition. *InfoRM J. UK Inst. Risk Management*, 11–12.

Hofstede, G. H. (1982). *Culture's consequences: International Differences in Work-related Values* Abridged ed. Newbury Park, CA: Sage Publications Inc.

Hofstede, G. H. (2001). *Culture's Consequences: Comparing Values, Behaviors, Institutions, and Organizations Across Nations.* 2nd ed. Thousand Oaks, CA: Sage Publications.

Hopkinson, M. (2000). The risk maturity model. *Risk Management Bull.* 5(4):25–29.

Hopkinson, M. (May 2000). Using risk maturity models. *Kluwer's Risk Management Briefing* 40(11):4-8.

Hulett, D. T. (2001). Key characteristics of a mature project risk organisation. In: *Proceedings of the 32nd Annual Project Management Institute Seminars, Symposium* (PMI 2001), presented in Nashville, USA, 5–7 November 2001.

Hulett, D. T. (2001). Key components of a mature risk management process. In: *Proceedings of the 4th European Project Management Conference* (PMI Europe 2001), presented in London, UK, 6–7 June 2001.

Hulett, D. T., Hillson, D. A., Kohl, R. (2002). Defining risk: a debate. *Cutter IT J.* 15(2):4–10.

Institution of Civil Engineers (ICE) and Faculty, Institute of Actuaries. (1997). *Risk Analysis, Management for Projects (RAMP)*. London: Thomas Telford.

Janis, I. (1972). *Victims of groupthink*. Boston: Houghton Mifflin.

Janis, I. (1982). *Groupthink: Psychological Studies of Policy Decisions and Fiascos*. Boston: Houghton Mifflin.

Kahneman, D., Slovic, P., Tversky, A. eds. (1986). *Judgement Under Uncertainty: Heuristics and Biases*. Cambridge, UK: Cambridge University Press.

Kaplan, S. (1996). *An Introduction to TRIZ*. Southfield, MI: Ideation International Inc.

Krimsky, S., Golding, D. (1992). *Social theories of risk*. Westport, CT: Praeger.

Lopes, L. L. (1987). Between hope and fear: the psychology of risk. *Adv. Exp. Social Psychol*. 20:255–295.

Marris, C., Langford, I. (September 1996). No cause for alarm, *New Scientist*, 28:36-39.

Marsh, J. G., Shapira, Z. (1987). Managerial perspectives on risk and risk-taking. *Management Sci*. 33(11):1404–1418.

McCray, G. E., Purvis, R. L., McCray, C. G. (2002). Project management under uncertainty: the impact of heuristics and biases. *Project Management J.* 33(1):49–57.

McGowan, C. (2000). Elements of risk: attitude measurement and the implications for senior management. MBA research thesis, University of Otago, Dunedin, New Zealand.

McKenna, S. (2001). Organisational complexity and perceptions of risk. *Risk Management: Internat. J.* 3(2):53–64.

Millward, L., Hopkins, L. (1997). Personal communication. Centre for Employee Research, Guildford, Surrey, UK.

Neal, R. A. (1995). Project definition: the soft-systems approach. *Int. J. Project Management* 13(1):5–9.

Norem, J. K. (2002). *The Positive Power of Negative Thinking*. New York: Basic Books.

Oldfield, A., Ocock, M. (1997). Managing project risks: the relevance of human factors. *Int. J. Project Business Risk Mgt*. 1(2):99–109.

Oldfield, A. (1998). The human factor in risk management. *Project* 10(10): 13–15.

Peterson, C. (2001). Assessing risk attitude for improved visibility to project risk. In: *Proceedings of the 4th European Project Management Conference* (PMI Europe 2001), presented in London, UK, 6–7 June 2001.

Pooley, R. (2001). International project teams—bridging the culture gap. In: *Proceedings of the Effective Project Management 2001 Conference*, held in London, UK, 30–31 October 2001.

Project Management Institute. (2000). *A Guide to the Project Management Body of Knowledge (PMBoK(r))* 2000 ed. Philadelphia: Project Management Institute.

Raz, T., Michael, E. (2001). Use and benefits of tools for project risk management. *Int J Project Management* 19(1):9–17.

Scarbrough, D., Gott, R., Hillson, D. A. (2003). Benchmarking organisational business risk management maturity: introducing the business risk management maturity model BRM3. In: *Proceedings of the International Association of Contract and Commercial Managers (IACCM) 2003 European conference*, held in London, 13–14 October 2003.

Schock-Smith, A. J. (2000). Risking a last stand. In: *Proceedings of the 31st Annual Project Management Institute Seminars, Symposium* (PMI 2000), presented in Houston, TX, USA, 7–16 September 2000.

Sennara, M., Hartman, F. (2002). Managing cultural risks on international projects. In: *Proceedings of the 33rd Annual Project Management Institute Seminars, Symposium* (PMI 2002), presented in San Antonio, USA, 7–8 October 2002.

Simon, P. W., Hillson, D. A., Newland, K. E., eds. *Project Risk Analysis, Management (PRAM) Guide.* High Wycombe, Bucks, UK: APM Group.

Slovic, P. (1987). Perception of risk. *Science* 236:280–285.

Steiner, C., Perry, P. (1997). *Achieving Emotional Literacy: A Personal Program to Increase Your Emotional Intelligence.* New York: Hearst Books.

Steiner, C., Perry, P. (2000). *Achieving Emotional Literacy.* 2nd ed. New York: Barnes, Noble.

Sutton, R. I. (2001). The weird rules of creativity. *Harvard Business Rev.* 79(8):94–103.

Trompenaars, F., Hampden-Turner, C. (1998). *Riding the Waves of Culture.* 2nd ed. New York: McGraw-Hill.

Tversky, A., Kahneman, D. (1974). Judgement under uncertainty: heuristics and biases. *Science* 185:1124–1131.

UK Office of Government Commerce (OGC). (2002). *Management of Risk—Guidance for Practitioners.* London: Stationery Office.

Webb, A. (1998). Some software applications to the problems of risk analysis: an empirical classification and review of some currently available products, Parts 1 and 2. *Int. J. Project Business Risk Management* 2:309–321, 401–422.

Webb, A. (2002). TRIZ—warm future for cold war survivor, Part 1. *Project Manager Today* 14(11):22.

Webb, A. (2002). TRIZ—warm future for cold war survivor, Part 2. *Project Manager Today* 15(1):16–18.

Whitney, D., Cooperrider, D. L. (1998). The appreciative inquiry summit: overview and applications. *Employment Relations Today* 25(2):17–28.

Yates, J. F., ed. (1992). *Risk-Taking Behaviour.* Chichester, UK: John Wiley.

10
Future Opportunities!

So the case has been made. It is undeniable that businesses and their projects have to operate in an uncertain environment. Not all uncertainty poses a risk however, but only that subset which has the potential to affect objectives. And within that subset there are uncertainties that if they occurred would be hurtful to project or business objectives, and there are others that would be helpful if they occurred. Both threats and opportunities must be managed proactively if the chance of project and business success is to be maximized. Since they are in essence two types of the same thing, namely uncertainties that if they occurred would affect one or more project objectives, threats and opportunities can be managed together via a joint process. The typical threat-focused risk management process is an obvious candidate for achieving this aim, and simple process extensions are possible to incorporate opportunity management alongside threats.

There are of course alternative solutions. An organization may choose to reserve its current risk management process for dealing exclusively with threats, and implement a separate approach to opportunity management distinct from the way threats are handled. Or threat-risk management and opportunity management might be combined in some overarching "uncertainty management," with subprocesses to deal with downside (risk) and upside (opportunity). These solutions can work, though they may not deliver all the benefits and efficiencies available from a combined process. Organizations must recognize that opportunities exist and that they must be managed proactively, otherwise the potential benefits they offer to projects and organizations will be lost, or at best only realized by chance. The important question for each business and project to answer is whether they have a structured way of dealing with opportunities, and if not, how they can implement that as quickly and effectively as possible. The relative costs of introducing and deploying a new process focused on opportunity management, or widening an existing risk management process, should be considered as part of that decision.

The central thesis of this book is that the most efficient and effective way to deal with opportunities is to utilize the existing familiar methods of risk management. Time will tell whether this recommendation finds favor with the project management community, risk management practitioners, and business at large. It is, however, incumbent on those promoting the inclusive position to make clear the implications of adopting this approach.

This final chapter addresses some of the issues that will require attention if risk management is to be broadened successfully to include opportunity. Although the case has been made that a wider scope for risk management is a logical and positive development, not all the pieces are yet in place to make the implementation effective and trouble-free. What is still needed to make it work in practice?

STANDARDS AND GUIDELINES

Official recognition by the professional bodies and standards organizations is an important first step for supporting the implementation of effective management of opportunity within the scope of risk management. While the most recent risk management standards do have inclusive definitions of "risk" that clearly mention opportunity or upside or positive impact as well as threats (see Table 1 and Fig. 1 of Chapter 2), the detailed processes are less clear. Many of these fail to follow through consistently on their initial declaration of intent. So although the definition of "risk" may include both opportunity and threat, standards still tend to talk about "risk mitigation" and the need to reduce risk exposure.

A key supporting requirement for a successful broadening of the risk process to include opportunities is therefore an unambiguous and explicit recognition of this position in the key standards and guidelines documents. Good progress is being made in some areas; for example, the risk management chapter in the forthcoming issue (2004) of the *Guide to the Project Management Body of Knowledge* (PMBoK) from the Project Management

Institute (PMI) is likely to be more explicit in its handling of upside risk, as will the next update (also expected in 2004) of the *Project Risk Analysis and Management (PRAM) Guide* from the UK Association for Project Management (APM). Other standards-issuing bodies should pay attention to this issue when revisions become due, and ensure that they come into line with latest thinking and practice.

If the leading professional bodies for project management and risk management indicate that an inclusive risk process is becoming generally accepted best practice, then this must be reflected in standards and guidelines, which should occupy a leadership position in driving best-practice developments into normal practice. There should be a mutually reinforcing relationship between thought leaders in the risk field and leading-edge risk practitioners, on one hand, and official risk management standards and guidelines, on the other. Where those at the frontiers of any discipline are pioneering new ways of achieving greater goals, they must be supported by others who consolidate and promulgate their findings to the majority. This is the role of standards and guidelines, to make best practice accessible and achievable to everyone, not just to those at the forefront of the field.

"Best practice" can be defined as "routine activities that lead to excellence," with the emphasis on the need for consistent implementation. Best practice does not mean undertaking special measures in particular circumstances, for example implementing opportunity management only on high-risk or strategically important projects. Instead best practice can be said to occur when the activities performed as part of the normal way of doing business naturally lead to excellence. Thus if proactive management of opportunities alongside threats in a common risk management process is defined to be best practice, implementing this approach consistently on all projects across the business will naturally result in successful projects, high-performing teams, increased business benefits, growth in reputation, and all the other attributes associated with excellence.

And if this approach to management of both threats and opportunities together genuinely represents best practice for all projects and business types, then it must be encapsulated in standards and guidelines for project management and risk management to make these benefits available to the wider project and business community.

PROCESS MODIFICATIONS

The most obvious area requiring changes is the risk management process, and Part II (Chapters 3–8) has addressed these in some detail. Generally speaking, the necessary process modifications are simple extensions of the techniques and methods that are familiar from the usual threat-focused approach to risk management. There are, however, a number of new techniques or novel applications of existing methods that are recommended for inclusion in the risk process to assist in identifying, assessing, and managing opportunities. Since some of these will be unfamiliar to most risk practitioners, project managers, and their teams, successful implementation will require development of training and user guides to support use of the new techniques.

Additional Recommendations

Novel additions to the risk management process suggested in previous chapters include the following:

Definition

- A clear statement in the Risk Management Plan that the risk process includes opportunities within its scope, and definition of risk acceptability thresholds for both threats and opportunities.

Risk Identification

- *SWOT Analysis*, ensuring that this proceeds to analysis rather than remaining as mere "SWOT identification," examining the influences of organizational factors (Strengths and Weaknesses) on project uncertainties (Opportunities and Threats), and considering the overall balance between upside and downside.
- *Constraints Analysis*, exploring the possibility of relaxing or removing constraints that have been imposed on the project, since absence of a constraint might create an opportunity to enhance the project or facilitate achievement of objectives.
- *Force Field Analysis*, explicitly exposing those driving factors that support achievement of project objectives and those restraining factors that hinder, then looking for opportunities to reduce restraints and reinforce drivers, as well as seeking threats that might strengthen restraints or weaken drivers.
- *Futures Thinking*, using scenario planning visualization techniques to explore alternative futures, including beneficial scenarios with significant opportunities, as well as less attractive scenarios containing high levels of threat.

- *Value Management*, a structured method for seeking to generate creative ideas for enhancing value by increasing upside and/or reducing downside, leading to identification of both opportunities and threats to project objectives.

Qualitative Risk Assessment

- Definition of project-specific impact scales to include *positive impacts* that would arise from the capture of opportunities, as well as the negative impacts of threats that might occur.
- Use of the *"mirror P-I Grid"* to depict both threats and opportunities, allowing the worst threats and best opportunities to be prioritized for further attention and action, and enabling the effects of risk responses to be visualized.
- Risk categorisation using a *Risk Breakdown Structure* (RBS) to identify common sources of threats and opportunities faced by the project, allowing management attention to be focused where it will be most effective.

Quantitative Risk Analysis

- Explicit inclusion of *opportunity information* when generating input data for a Monte Carlo risk model, for example reflecting the potential upside impact of opportunities in the optimistic or best-case values.
- Use of *stochastic branches* to model key opportunities in the risk model, with alternative logic reflecting the possibility of exploiting opportunities to work faster or cheaper, and also including null paths showing the opportunity to remove tasks, or negative lags to indicate the possibility of starting some tasks earlier than planned.

Risk Response Planning

- Explicit recognition of strategy options for opportunities, including aggressive *exploitation* seeking to capture an opportunity and make it definitely happen, risk *sharing* to involve other stakeholders in managing an opportunity, *enhancement* actions to make the opportunity more likely or more beneficial, and *acceptance* strategies taking no specific action except perhaps allocating suitable contingency.

- Development of *specific actions* to implement opportunity strategies, to ensure that the good intentions of the response planning phase are translated into reality.
- Considering the possibility of *secondary opportunities* arising from implementation of responses to identified risks.

Monitoring, Control, and Review

- Modifying outputs from the risk process to include and communicate *opportunity information*, including the Risk Register, risk reports, metrics, etc.
- Ensuring an effective *interface to a benefits management process*, so that the impacts of opportunities that have been successfully captured can be exploited fully, and the promised benefits are realized for the project and the organization, even though they were initially unexpected and unplanned.

Each of these novel process extensions needs to be defined and explained to participants in the risk process, so that they are aware of what each technique offers, its strengths and weaknesses, and practical implications of its use. This is likely to involve a degree of skills training so that new opportunity-focused techniques can be used effectively.

Procedure development will also be required so that new opportunity techniques interface seamlessly with existing threat-focused techniques and other project management and business processes. It is important that the opportunity management element of the risk process should be seen as a fully integrated part of the whole, rather than an optional extra to be implemented in special cases. The goal is for participants in the risk management process to implement all the necessary risk techniques equally and naturally, without considering whether a particular technique is relevant to threats or opportunities.

IMPLICATIONS FOR TOOLS

An essential corollary of implementing an inclusive risk management process covering both threats and opportunities is for the supporting risk tool set to provide the necessary functionality. Like standards and guidelines, this is another area where the market will naturally follow development of best practice. As leading thinkers and practitioners in risk management make the case for a broadening of the risk process, so increasing numbers of organizations and projects will wish to implement the approach rec-ommended as best practice. This will expose functionality shortfalls in currently available supporting tools, and create a demand for new develop-ments to which vendors will respond.

It is unlikely that tool vendors will take the lead to introduce spec-ulatively the functionality required to support opportunity management within the risk process unless there is a demand from users, although some forward-looking software companies have already noticed the growing trend and have moved ahead to offer the needed support. Such proactive companies will of course be in a better position than their reactive rivals to exploit the new market that opportunity management will create when it is adopted more widely.

The primary candidate for enhancement is the ability to deal with opportunities within the qualitative Risk Assessment tool set. Typical tools currently available include risk register software, based on an underlying database structure, many of which also produce analytical outputs such as prioritized lists of "top risks," various risk distributions, risk metrics, and Probability-Impact (P-I) Grids. These tools will need to be modified to accept opportunity data and allow opportunities to be analyzed and re-ported alongside threats. Coding and numbering structures must be ca-pable of handling both opportunities and threats, and distinguishing them where necessary for separate analysis and reporting. Outputs must present opportunity and threat data in a variety of formats, allowing the overall risk exposure comprising both upside and downside to be assessed and reported. This is likely to include the "mirror-format" double P-I Grid, modifications to risk metric calculations, and SWOT-type outputs.

Some existing Risk Identification tools will also require modification to support the inclusion of opportunities, particularly brainstorm facilita-tion support tools, and risk identification checklists (both paper-based and automated).

Some of the novel techniques listed above will require development of new tools to provide automated support, such as SWOT Analysis, Con-straints Analysis, Force Field Analysis, or Futures Thinking.

Other existing tools can be used without the need for significant changes, for example some Monte Carlo simulation software packages for quantitative Risk Analysis, which already provide all the necessary func-tionality to allow opportunity-based data to be included in the risk model (although the capability of some Monte Carlo tools might require enhance-ment, for example to deal with negative lags, or representation of oppor-tunities on Tornado charts).

The widespread adoption of opportunity management within the standard risk process will present a significant new market for risk tool vendors. The current marketplace is fairly well served with a variety of existing tools covering all the main areas where support is required. Vendors who identify the current trend in risk management thinking to include the upside will be able to exploit this new opportunity for growth and expansion, by matching the trend and providing tools to support the new requirement. As always, the greatest advantage will be gained by those prepared to take risk, especially when that risk represents a new opportunity with significant potential benefits.

IMPLEMENTING EFFECTIVE OPPORTUNITY MANAGEMENT

Once the standards and guidelines, processes and procedures, and supporting tools are all in place, there will be no excuse for not performing opportunity management as an integral part of the risk process. The biggest single remaining barrier to successful implementation is likely to be the prevailing risk attitudes of individuals and organizations, as discussed in Chapter 9.

The proverb rightly says, "You can lead a horse to water but you can't make it drink." This applies equally to a broad approach to risk management that includes opportunities. Policies, procedures, and processes may be issued, with the necessary infrastructure to enable opportunities to be identified, assessed, and managed, but it is still people who perform the activities within the risk process. If individual risk attitudes and/or corporate risk culture are not conducive to opportunity management, provision of tools and processes will not suffice. Even attendance on training courses to raise awareness of the benefits of proactive management of opportunities as part of risk management can fail to produce the necessary motivation to overcome attitudinal and cultural barriers.

In the final analysis, standards, processes, and tools are merely the scaffolding that supports the main structure—they are not the building itself. Management of project risk, whether threats or opportunities or both, is an activity performed by people within organizations. This is why it is so important to understand individual risk attitudes and corporate risk culture, and to modify these through determined and conscious actions to create the required environment where opportunity is recognized and sought, even in the face of considerable uncertainty. Emotional Intelligence profiling tools can be used to raise self-awareness and encourage development of modified personal attitudes and more risk-mature corporate cultures. It is also necessary to develop techniques for diagnosing the operation of the various heuristics that affect decisions made under conditions of uncertainty, so that any resulting bias can be identified, scoped, and corrected for.

Some additional tools would be useful here, since this is probably the area requiring most development for organizations implementing a structured approach to risk management. Very few have any formal methods in place for assessing or modifying individual attitudes or corporate culture. The benefits of consciously building risk-balanced teams containing a spread of risk attitudes cannot be achieved without reliable diagnostic tools that are easy to apply and interpret. Similarly, the organizational development approaches that are best suited to encouraging opportunity thinking are not well established, and techniques such as TRIZ, Appreciative Inquiry, and emotional literacy methods should be explored and applied more widely.

PRACTICAL CONSEQUENCES

In the final resort, inclusion of opportunity management as an integral part of the risk process will become widespread and accepted only if there are demonstrable benefits. In theory there are many such benefits, but these must be seen in practice if busy project managers and their teams or resource-stretched businesses are to accept the need for additional effort. It has been rightly said that "In theory there is no difference between theory and practice, but in practice there is!" So what practical benefits might be expected to arise from an inclusive approach to risk management that addresses both threats and opportunities?

> *No new process.* The use of a common process for managing both threats and opportunities ensures maximum efficiency, with the same activities being used for both types of risk. There is no need to develop, introduce, and maintain a separate opportunity management process, since the one common process will suffice.
>
> *Extension from the familiar techniques.* Most of the techniques currently used to manage threats can be adapted with only minor

changes to deal with opportunities. Any additional techniques are also likely to be familiar since they are drawn from related disciplines.

Minimized additional overhead. The infrastructure in place for existing threat-based risk management should be able to support a broadened process without requiring significant modifications.

Minimal additional training. Since the common process is building on the familiar, there is no need to train staff in the process, tools, and techniques of including opportunities, apart from ensuring awareness of upside risk, and providing training for new techniques.

Enhanced benefits. Including opportunity management within the risk process ensures that effort is expended to look for opportunities and exploit them proactively. This means that opportunities that might have been missed can be tackled, and some of them will be captured.

Cost-effectiveness (double "bangs per buck"). The use of a single process to achieve proactive management of both threats and opportunities will result in avoidance or minimization of problems as well as exploitation and maximization of benefits.

Building on existing stakeholder commitment. Project teams and other project stakeholders are already used to thinking about downside uncertainty and managing it proactively through the risk process. The extension to include opportunity is a natural progression.

Better contingency management. Inclusion of potential upside impacts as well as the downside means that contingency calculations are likely to be more realistic, taking account of both "roundabouts and swings."

Increased team motivation. Encouraging people to think creatively about ways to work better, simpler, faster, more effectively, etc. is a great motivator, and teams will enjoy looking for opportunities and making them happen.

Enhanced professionalism. Clients who see organizations and project teams working to maximize the benefits on their project will be impressed at the display of professionalism, leading to increased reputation and business growth.

Improved chances of project success. As opportunities are identified and captured, so projects will gain the associated benefits that would otherwise have been missed, leading to more successful projects that achieve (or even exceed) their objectives.

One of the best ways to encourage widespread adoption by businesses and projects of management of opportunity within the risk process is for those organizations at the leading edge already implementing this approach to demonstrate the benefits. When others see the advantages gained from the integrated threat + opportunity approach, increasing numbers of them will choose to follow the trailblazers.

This requires organizations to be prepared to measure the effectiveness of their combined risk management process, recording costs and benefits, and attempting to calculate a Return On Investment (ROI) for risk management (though this is not a simple matter). It also means that they must share their findings with the wider business and project management community, allowing others to benchmark themselves against the leaders. Open disclosure is required to allow learning from the experience of others, including failures as well as successes.

FINAL THOUGHTS

Opportunities exist and must be managed by organizations and projects wishing to adopt a professional approach to business and project management. Missing an achievable opportunity that could have delivered tangible benefits is as bad as allowing an avoidable threat to mature into a problem. Everyone will agree that both threats and opportunities must be tackled proactively. The question addressed by this book is whether the most effective way to accomplish this is by using a combined approach based on extension of the familiar risk management process. The answer reached in the preceding chapters is an emphatic affirmative. Effective opportunity management for projects and businesses can be achieved by building on tried and tested risk management techniques and methods. There is no need to invent something new—all that is required is some simple modifications of what already exists.

Although different solutions can be found to the requirement for proactive management of opportunities, there seems to be a trend in some organizations toward using a broadened risk management process to address both threats and opportunities. This is supported by the most recent risk management standards and guidelines issued by professional bodies and others, whose definitions of "risk" include both upside and downside. This book has demonstrated the implications of this trend, in terms of definitions, process, Critical Success Factors, individual risk attitudes, and risk-mature corporate culture.

Those organizations choosing to follow the route recommended in this book will find in its pages a set of guidelines and recommendations to make a smooth transition from threat-only risk management into an inclusive approach to managing upside opportunities together with downside threats. In fact, this represents a significant opportunity, with clear potential benefits. Of course there are uncertainties associated with embarking on this path. Applying the approach described here will enable organizations and their projects to gain the many advantages offered by integrating opportunity management within the traditional risk process. Significant benefits await those prepared to take the risk and capture the opportunities!

Appendix: Results from a Survey Exploring Definitions

THE ISSUE

There is no doubt that projects, like everything else in life, are subject to uncertainty. It is also clear that some of that uncertainty might be harmful if it came to pass (i.e., "threat"), whereas other uncertainties might assist in achieving objectives (i.e., "opportunity"). The issue is whether both types of uncertainty could or should be included in the definition of "risk," and whether both could or should be handled by a common "risk management process." There is currently an active debate among risk management practitioners about the definition of risk (see, for example, Chapter 2 of this book; also Hillson, 2002a, 2002b; Hulett and Hillson, 2002; Kohl, 2002). This is not simply an idle discussion about terminology. It has clear implications for the risk management process, because encompassing both opportunities and threats within a single definition of risk is a clear statement of intent, recognizing that both are equally important influences over business and project success, and both need managing proactively. Opportunities and threats are not qualitatively different in nature, since both involve uncertainty, which has the potential to affect objectives. As a result, both can be handled by the same process, although some modifications may be required to the standard risk management approach to deal effectively with opportunities (as detailed in Chapters 3–8; also in Hillson, 2001, 2002a).

Some risk practitioners (for example, Kohl, 2002) object to use of a broad definition of risk on the basis that common current usage of the word "risk" is exclusively negative, stating that this tendency cannot be overcome and should not be changed within the risk management or project management fields. This objection can be met by two arguments (see Hulett and Hillson, 2002, for a more complete discussion). First, words change their common-use meanings over time; and second, it is quite acceptable for a profession to adopt a technical usage that differs from the layman's definition. In addition, both opportunities and threats can be handled by a common "risk management" process, with clear advantages. Extending the existing threat-based process to cover opportunities will mean that there is no need to introduce a new process, with all the resistance that is likely to be encountered when another overhead is added. There is clear synergy in extending the same process to cover both types of uncertainty—if the management team are already setting aside time and effort to deal with threats, why not use the same time to identify and proactively manage opportunities as well?

It is clear that the risk management process as commonly implemented does not include a structured framework for proactively addressing opportunities, since it focuses almost entirely on threats. If a broadened definition of risk is used that includes both opportunities and threats, and if an extended approach is implemented to address both together, the organization will be able to take full advantage of those uncertainties with potential upside impact. The alternative is likely to be a failure to implement proactive opportunity management strategies, which will guarantee that only half of the benefits of risk management can be achieved.

The review of published risk management standards and guidelines in Chapter 2 indicates that a wide variety of definitions of risk are in current use, though there appears to be some momentum toward a broadened approach that includes both upside opportunities and downside threats. This remains an issue of active debate among risk management professionals, so it is of interest to determine the extent to which current practice is following the standards documentation. With such a range of different definitions being offered by standards and guidelines, what are risk practitioners actually doing in practice? The survey discussed below was designed to establish evidence of current practice among risk practitioners and determine the risk definitions actually used in the risk management community.

SURVEY PROCESS

The Risk Management Specific Interest Group of the Project Management Institute (PMI Risk SIG) and the Risk Management Working Group of the International Council On Systems Engineering (INCOSE RMWG) are international practitioner networks representing significant groupings of professionals active in project risk management. Leading members of these two bodies therefore decided to poll members and others to gauge international opinion on the definition question. A simple questionnaire was designed (see Table 1 for survey questions) to explore what definitions were currently in use by the organizations represented by respondents, as well as the personal perspectives of respondents.

The questionnaire was distributed by e-mail to members of several professional bodies to which risk management practitioners belong. This included the PMI Risk SIG, INCOSE RMWG, the Risk SIG of the UK Association for Project Management (APM), the UK Institute of Risk Management (IRM), and the Global Association of Risk Professionals (GARP). The total number of members from these groups who were invited to participate by e-mail in the survey was approximately 2000 (note that the actual total membership of these groups is much higher, but not all could be polled by e-mail). The survey was distributed on 1 September 2001, with a return date of 31 October 2001.

SURVEY RAW RESULTS

Respondents

A total of 186 responses were received, representing a response rate of about 10%. This rate is typical of surveys conducted by e-mail, and provides some assurance that the results might be representative of the general views of risk management practitioners in the groups surveyed.

Respondents came from a wide range of industries, with high representation from consultants (32% of respondents) and information technology (IT) companies (29%). There were also significant numbers of responses from people working in the communications industry (9%) and government agencies (8%).

Most of the respondents claimed to be members of the PMI Risk SIG (73%), and 15% stated that they belonged to an IPMA-affiliated body such as APM. Only small numbers reported membership of INCOSE

Table 1   Risk Definition Survey Questions

Q1. Which of the following definitions of risk is closest to that used by your organization?	a. ___ Risk is an uncertain event or condition which, if it occurs, would have an undefined or unknown impact on achievement of objectives.
	b. ___ Risk is an uncertain event or condition which, if it occurs, would have a negative impact on achievement of objectives (threat).
	c. ___ Risk is an uncertain event or condition which, if it occurs, would have a negative or positive impact on achievement of objectives (threat or opportunity).
	d. ___ Some other definition (please state)
Q2. Which of the following best describes your organization's approach to risk management?	a. ___ The risk management process aims to manage potential negative impacts on objectives (i.e., threats only). There is no process for explicit handling of opportunities.
	b. ___ The risk management process aims to manage potential negative impacts on objectives (i.e., threats only). Opportunities are handled via a separate process that is not an integrated part of risk management.
	c. ___ The risk management process aims to manage both threats and opportunities in a common (integrated) process.
	d. ___ Some other approach not covered by the above (please state).
Q3. Which of the definitions in Question 1 best reflect your own preferred definition of risk?	a. ___ Risk is an uncertain event or condition which, if it occurs, would have an undefined or unknown impact on achievement of objectives.
	b. ___ Risk is an uncertain event or condition which, if it occurs, would have a negative impact on achievement of objectives (threat).
	c. ___ Risk is an uncertain event or condition which, if it occurs, would have a negative or positive impact on achievement of objectives (threat or opportunity).
	d. ___ Some other definition (please state)
Q4. Would you (or do you) support a general change in the definition of risk to include both threats and opportunities?	a. ___ Yes
	b. ___ No
	c. ___ Don't know
	d. ___ Don't care

Table 1 Continued

Q5. Would you like to receive a summary of the survey results?	___ Yes : my e-mail address is _____ ___ No

Q6. Indicate the industry sector/organization type in which you work.

a. ___ Manufacturing
b. ___ Research and Development
c. ___ Aerospace
d. ___ Defense
e. ___ Information Technology
f. ___ Construction
g. ___ Transportation
h. ___ Finance
i. ___ Government or Public Agency
j. ___ Consulting
k. ___ Energy
l. ___ Not-for-Profit
m. ___ Environmental
n. ___ Communications
o. ___ Agriculture
p. ___ Medical
q. ___ Other (please state)

Q7. Comments (optional)

Q8. Personal details (optional)

Name (optional) _____

Organization (optional) _____

Membership in Risk Management-related professional bodies (check all that apply) :

___ Project Management Institute (PMI)

___ International Council On Systems Engineering (INCOSE)

___ IPMA-affiliated body (e.g., UK Association for Project Management)

___ UK Institute of Risk Management (IRM)

___ Global Association of Risk Professional (GARP)

___ Other (please state)

RMWG, IRM, or GARP, and 10% belonged to some other body. However, 15% of respondents were members of more than one professional body so there is some overlap between these results.

Question 1: Organizational Definition of Risk

Respondents were asked in the first question (see Q1 in Table 1) to choose a definition of risk closest to that used by their organizations, with four options given, as follows :

    a.   Risk is an uncertain event or condition which, if it occurs, would have an undefined or unknown impact on achievement of objectives.

    b.   Risk is an uncertain event or condition which, if it occurs, would have a negative impact on achievement of objectives (threat).

    c.   Risk is an uncertain event or condition which, if it occurs, would have a negative or positive impact on achievement of objectives (threat or opportunity).

    d.   Some other definition.

The results (in Fig. 1) show that about half of the organizations with which respondents work (54%) use a definition of risk with exclusively negative effects, i.e., a threat-only definition. About a third (34%) use a broader definition of risk that includes both threats and opportunities.

Question 2: Organizational Approach to Risk Management

The second question (Q2) asked about the approach to risk management within respondents' organizations, to determine whether a threat-only risk process was used, and how opportunities were managed (if at all). Options for this question were:

    a.   The risk management process aims to manage potential negative impacts on objectives (i.e., threats only). There is no process for explicit handling of opportunities.

    b.   The risk management process aims to manage potential negative impacts on objectives (i.e., threats only). Opportunities are handled via a separate process that is not an integrated part of risk management.

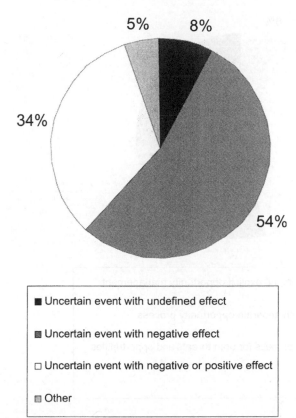

5%   8%

34%

54%

- ■ Uncertain event with undefined effect

- ▨ Uncertain event with negative effect

- □ Uncertain event with negative or positive effect

- ▥ Other

Figure 1 Q1: Organizational definition of risk.

    c. The risk management process aims to manage both threats and opportunities in a common (integrated) process.
    d. Some other approach not covered by the above.

Results are in given Figure 2, which shows that over half of the organizations represented (54%) use the risk management process only to manage threats. These are split almost equally between those who only have a threat-focused risk process with no explicit opportunity management (26%), and those who use a separate process for opportunity management in addition to the threat-based risk process (28%).

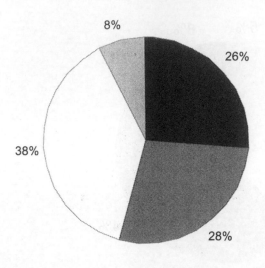

Figure 2   Q2: Organizational approach to risk management.

Over a third of organizations (38%) have a common risk management process that is used to manage both threats and opportunities in an integrated fashion.

Question 3: Personal Definition of Risk

People invited to complete this questionnaire were members of professional bodies that specialize in risk management. It is therefore possible that individual risk practitioners might hold a different view of risk from the organizations for which they work. The third question (Q3) was included to test this possibility. Individuals were asked which of the definitions of risk stated in the earlier question best reflected their own preferred definition of risk. The results are shown in Figure 3. The

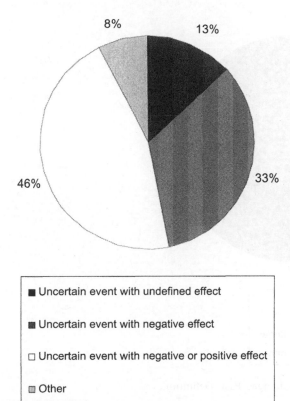

8%   13%

46%   33%

■ Uncertain event with undefined effect

■ Uncertain event with negative effect

□ Uncertain event with negative or positive effect

▨ Other

Figure 3   Q3: Personal definition of risk.

distribution of replies to this question is the reverse of that reported for organizations, with almost half of the respondents (46%) stating that they prefer to use a broader definition of risk that includes both threat and opportunity. A third (33%) use a threat-only definition. It is interesting to speculate on whether this difference of opinion between the individual risk practitioner and his organization might cause any difficulties in performing the risk process.

Question 4: Support for Changed Risk Definition

One of the main purposes of the survey was to identify whether risk practitioners would support a general change in the definition of risk to in-

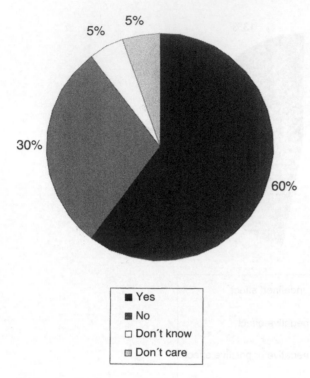

Figure 4 Q4: Support for Changed Risk Definition.

clude opportunities as well as threats, since this is the topic of current debate among the risk community. The fourth question (Q4) asked directly whether respondents would (or already did) support such a change. Figure 4 shows that 60% support a move toward a broader definition, with 30% opposed to change.

ANALYSIS OF RESULTS

The preceding section presents the raw data from the survey, giving answers to each question as percentages of the total response. More detailed analysis, however, allows further conclusions to be drawn about current perspectives on the nature of risk and the use of risk management. These are explored below.

Industry-Specific Views of Risk

Comparing responses to Questions 1 and 6 (see Table 1) indicates whether particular groups of respondents hold a shared view of risk. Since a large proportion of replies were received from consultants (selecting Q6j) or those working in the IT sector (Q6e), it is possible to see whether these respondents tend to hold threat-only definition of risk or a broader definition.

Among IT respondents, 55% replied that they defined risk only in terms of threats (Q1b), with 33% using a definition including both threat and opportunity. These proportions match closely the overall distribution from all responses (Q1b = 54%, Q1c = 34%). The split among consultants is rather different however, with 48% of consultants defining risk exclusively as threats and 40% using the broader definition.

Although numbers of replies from other sectors were smaller, it may be significant that about 60% of those from the communications and government agency sectors saw risk as only negative. The sector with the highest percentage using a "threat + opportunity" definition was transportation (66%) though the number of replies was not statistically significant.

Organizational vs. Personal View of Risk

Comparing the results of Question 1 and Question 3 (see Figs. 1 and 3) reveals an interesting juxtaposition between the definition of risk used by organizations and that used by individual respondents. About half of the organizations use a "threat-only" definition (selecting Q1b) and a third use a broader definition of "threat + opportunity" (Q1c). However, for individual risk practitioners, about half define risk to include both threat and opportunity (Q3c), while a third use "risk" to refer only to a threat (Q3b). There are a number of possible explanations for this finding, and the available data cannot determine between them.

- Perhaps individual risk practitioners hold a more current view of risk than organizations, since it is their area of speciality. This might lead them to have adopted the wider definition including opportunity ahead of their organizations who are still using the older, more traditional negative definition.
- Another possibility is that those individuals responding to this survey were a self-selecting nonrepresentative group who felt more strongly about including opportunity in the risk definition than people or organizations at large. The predominance of project risk management practitioners may also have skewed the survey result,

since other groups of risk specialists may prefer to regard risk as only covering threats (e.g., those working in financial risk, insurance, actuarial risk, health and safety risk, etc.).

• Data quality must be considered when interpreting these (and other) results, since it is possible that respondents may be misrepresenting their organization's approach to risk management, whether consciously or not.

• Finally, it is possible that the difference between organizational and individual views may represent a time-lag effect, since individuals can respond more quickly than organizations to changes in definitions, standards, best-practice, or leading-edge.

Organizational Consistency

Replies to Questions 1 and 2 indicate a high degree of internal consistency in organizations in their approach to risk management. Q1 describes the definition of risk used in the organization, with Q2 detailing the approach to risk management.

A third of organizations (62 replies to Q1c, 34%) use an inclusive definition including both threats and opportunities, and about a third (71 replies to Q2c, 38%) have an integrated common process for managing both threats and opportunities together. Analysis indicates that these are largely the same organizations (with 50 responses in common), revealing a welcome consistency of approach. In other words, most (80%) of the organizations that use an inclusive risk definition also have an integrated risk process.

Organizations that define risk exclusively in terms of threats (i.e., 99 replies, or 54%, answering Q1b) are also consistent, tending to reserve their risk management process for threat management. These are almost equally split between those who have no explicit opportunity management process (39 also answered Q2a), and those who use a separate process for managing opportunities (42 answered Q2b). Only 20 respondents answered either Q2a or Q2b who did not also answer Q1b. In other words, most (80%) of the organizations with a negative risk definition also have a threat-focused risk process.

Appetite for Change

Answers to Question 3 reveal that 13% of individuals use a definition of risk with a neutral (i.e., undefined or unknown) impact (Q3a), while 33% use a threat-only definition (Q3b), and 46% use a broader definition including opportunity (Q3c).

Responses to Question 4 shows that 60% of respondents would support a general change in the definition of risk to include both threats and opportunities (Q4a), with 30% opposed to change (Q4b).

Analysis indicates that those respondents supporting change are largely the ones who already use either a wider risk definition or a neutral definition. Of the 112 individuals supporting change, 71% already define risk to include both threat and opportunity, and an additional 16% use a neutral definition.

On the other hand, people whose personal definition of risk is exclusively negative tend to oppose change: 72% of those with a threat-only definition of risk oppose change, and 82% of those opposed to a broader definition hold a threat-only definition.

This might indicate division of the risk community into traditionalists who hold and wish to preserve the view that risk is synonymous with threat, and progressives who already take a non-traditional view and wish to see it more widely accepted.

Organizational Membership

Analysis of organizational membership (see Question 8 in Table 1) reveals that members of the PMI Risk SIG hold mixed views on the definition question, with 59% supporting change and 32% opposing it. This is somewhat surprising, since PMI Risk SIG members might be expected to have been influenced by the risk management chapter in the 2000 edition of the PMI's *Guide to the Project Management Body of Knowledge* (Project Management Institute, 2000), which uses a broader definition of risk including both threat and opportunity. The other main group of respondents (15%) belonged to IPMA-affiliated bodies (largely the UK Association for Project Management), and these strongly supported change, with 82% in support of a wider risk definition and only 14% opposed.

The overwhelming majority of responses came from the project risk management community (88% belong to PMI and/or APM), and this may have skewed the results of the survey, since it is possible that they may not be representative of the body of risk practitioners as a whole.

CONCLUSIONS AND FURTHER WORK

This short limited survey has produced some interesting and significant data relevant to the current debate over the meaning that should be given to the term "risk." Survey data show that over half of organizations and a third of individuals currently use a definition of risk that is just negative, compared to a third of organizations and nearly half of individuals who use a wider definition including both threats and opportunities.

But the survey also indicates that the majority of respondents (60%) support changing the definition to be more inclusive. The results also suggest that many organizations do in fact use a common process to manage both threats and opportunities (38%), compared to 26% whose risk process only manages threats, and 28% who have two separate processes for threat-risks and opportunities.

These results can be interpreted to mean that there is some pressure toward an expanded definition of risk, and that such a change would reflect current practice. However, further data are required to test this tentative preliminary conclusion, drawing on the opinions of a wider constituency of risk practitioners. In particular, the general applicability of the results may be compromised by the concentration of responses from members of the project risk management community, and from the consultancy and IT sectors. It is therefore recommended that other practitioner bodies and other industry sectors should be encouraged to contribute to this or a similar survey, supplementing the current data and allowing stronger conclusions to be drawn.

Published risk management standards and guidelines offer a wide variety of different definitions for the term "risk," including some that are wholly negative, equating "risk" with "threat," others that use a neutral definition where the effect of risk is undefined, and a third group where "risk" is defined as including both upside opportunity and downside threat. The results of the survey reported here indicate that risk management practitioners also follow a range of different approaches. There is, however, a significant proportion of organizations and individuals who recognize both upside and downside risk, and there appears to be support for a change in the definition of risk to include both threats and opportunities.

ACKNOWLEDGMENTS

The survey reported here could not have been undertaken without the active participation and involvement of the following colleagues, who are all leading members of the participating groups, namely PMI Risk SIG and/or INCOSE RMWG:

    David Hall (SRS Information Services)
    Dr. David Hulett (Hulett and Associates)

Barney Roberts (Futron Corporation)
William Seeger (SAIC)

The author extends his sincere gratitude and appreciation to them for their input into defining and conducting the risk definition survey, and for their comments on the analysis and interpretation of survey data.

REFERENCES

Hillson, D. A. (2001). Effective strategies for exploiting opportunities. In: *Proceedings of the 32nd Annual Project Management Institute Seminars & Symposium* (PMI 2001), Nashville USA, 5–7 November 2001.

Hillson, D. A. (2002a). Extending the risk process to manage opportunities. *Int. J. Project Management* 20(3):235–240.

Hillson, D. A. (April 2002b). What is risk? Towards a common definition. *InfoRM, J. UK Inst. Risk Management* (April):11–12.

Hulett, D. T., Hillson, D. A. (2002). Defining risk: a debate—project risk includes opportunities. *Cutter IT J.* 15(2):4–10.

Kohl, R. J. (2002). Defining risk : a debate — if it ain't broke, don't fix it. *Cutter IT J.* 15(2):4–10.

Project Management Institute (2000). *A Guide to the Project Management Body of Knowledge (PMBoK®)*, 2000 ed. Philadelphia: Project Management Institute.

References and Further Reading

Adams, J. (1995). *Risk: The Policy Implications of Risk Compensation and Plural Rationalities*. London: UCL Press.

Adams, J. (1998). The risk thermostat: risk management is a balancing act. *Project* 10(8):18–19.

Akintoye, A. S., MacLeod, M. J. (1997). Risk analysis and management in construction. *Int. J. Project Management* 15(1):31–38.

Ansell, J., Wharton, F. (1992). *Risk Analysis, Assessment and Management*. Chichester, UK: John Wiley.

Artto, K. A., Kahkohnen, K., Pitkanen, P. J. (2000). *Unknown Soldier Revisited: A Story of Risk Management*. Finland, Helsinki: PMA.

Bailey, V., Shipway, M., Hillson, D. A. (1999). Millenium meltdown: surviving the immoveable deadline. *Project Achievement* (1): 33–34.

Barber, R. B., Burns, R. (2002). A system approach to risk management. In: *Proceedings of the ANZSYS 2002 conference*, held in Gold Coast, Australia, December 2002.

Bartlett, J. (2002). *Managing risk for projects and programmes: A Risk Handbook*. Hook, Hampshire, UK: Project Manager Today Publications.

Berstein, P. L. (1996). *Against the Gods: the Remarkable Story of Risk*. Chichester, UK: John Wiley.

Boehm, B. W. (1989). *Software Risk Management*. Los Alamitos, CA: IEEE Computer Society Press.

Boley, T. M. (1999). Enterprise-wide approach to managing organisational project risks. *PMI Risk SIG Newsletter* 1(4):1–4.

Boothroyd, C. E., Emmett, J. (1996). *Risk Management—A Practical Guide for Construction Professionals.* London: Witherby.

Boothroyd, C. E. (2000). Risky business. *Project* 13(3):24–26.

Borge, D. (2001). *The Book of Risk.* New York: John Wiley.

Boyce, T. (1995). *Commercial Risk Management.* London: Thorogood.

Browning, T. R. (1999). Sources of schedule risk in complex system development. *Systems Engineering* 2(3):129–142.

Browning, T. R., Deyst, J. J., Eppinger, S. D., Whitney, D. E. (2002). Adding value in product development by creating information and reducing risk. *IEEE Trans. Engineering Management* 49(4):443–458.

Burgess, M. (1999). Living dangerously: the complex science of risk: Channel 4 Television, London UK.

Carter, B., Hancock, T., Morin, J.-M., Robins, N. (1994). *Introducing RISKMAN.* Oxford, UK: NCC Blackwell.

CCTA. (1993). *Introduction to the Management of Risk.* London: HMSO.

CCTA. (1994). *Management of Project Risk.* London: HMSO.

CCTA. (1995). *Management of Programme Risk.* London: HMSO.

Chapman, C. B., Ward, S. C. (1997). *Project Risk Management: Processes, Techniques and Insights.* Chichester, UK: John Wiley.

Chapman, C. B., Ward, S. C. (2000). Estimation and evaluation of uncertainty—a minimalist first-pass approach. *Int. J. Project Management* 18(6):369–383.

Chapman, C. B., Ward, S. C. (2002). *Managing Project Risk and Uncertainty.* Chichester, UK: John Wiley.

Chapman, C. B. (1997). Project risk analysis and management—PRAM, the generic process. *Int. J. Project Management* 15(5):273–281.

Chapman, R. J. (2001). The controlling influences on effective risk identification and assessment for construction design management. *Int. J. Project Management* 19(3):147–160.

Charette, R. N., Edrich, C. (2001). *Implementing Risk Management Best Practices.* Arlington MA: Cutter Information Corp.

Charette, R. N. (1989). *Software Engineering Risk Analysis and Management.* New York: McGraw-Hill Education.

Charette, R. N. (1990). *Applications Strategies for Risk Analysis.* New York: McGraw-Hill Education.

Charette, R. N. (2002). *The State of Risk Management 2002: Hype or Reality?* Arlington, MA: Cutter Information Corp.

Chicken, J. C., Posner, T. (1998). *The Philosophy of Risk.* London: Thomas Telford.

Cooke-Davies, T. (1998). Can we afford to skimp on risk management? *Project Manager Today* 10(9):12–15.

Couillard, J. (1995). The role of project risk in determining project management approach. *Project Management J.* 26(4):3–15.

Courtney, H., Kirkland, J., Vignerie, P. (1997). Strategy under uncertainty. *Harvard Business Rev.* 75(6):66–79.

de Cano, A., de la Cruz, M. P. (1998). The past, present and future of project risk management. *Int. J. Project Business Risk Management* 2(4):361–387.

Dembo, R. S., Freeman, A. (1998). *Seeing Tomorrow—Rewriting the Rules of Risk*. New York: John Wiley.

Dorofee, A. J., et al. (1996). *Continuous Risk Management Guidebook*. Pittsburg, PA: Software Engineering Institute, Carnegie Mellon University.

Edwards, L. (1995). *Practical Risk Management in the Construction Industry*. London: Thomas Telford.

Engineering Council. (1995). *Guidelines on Risk Issues*. London: EC.

Flyvbjerg, B., Bruzelius, N., Rothergatter, W. (2003). *Megaprojects and Risk: An Anatomy of Ambition*. Cambridge, UK: Cambridge University Press.

Giddens, A. (1999). *Runaway World: How Globalisation Is Reshaping Our Lives*. Chapter 2: Risk. London: Profile Books. [Based on the 1999 BBC Reith Lectures, see http://news.bbc.co.uk/hi/english/static/events/reith99/week2/week2.htm.]

Godfrey, P. S. (1995). Control of risk: a guide to the systematic management of risk from construction. CIRIA under reference FR/CP/32. London: CIRIA.

Grey, S. (1995). *Practical Risk Assessment for Project Management*. Chichester, UK: John Wiley.

Hall, E. M. (1998). *Managing Risk—Methods for Software System Development*. Reading, MA: Addison Wesley Longman.

Hall, E. M. (1999). Risk management return on investment. *Systems Engineering* 2(3):177–180.

Hillson, D. A. (1995). Mission Possible. *Cost Plus (J. Assoc. Cost Engineers)* (June): 10–12.

Hillson, D. A. (1996). Use of risk analysis to determine MIS development strategy. In: *Management of Risk: Case Studies*. Vol. 2. London: HMSO.

Hillson, D. A. (January 1998). Managing risk. *IEEE Rev.* 44(1):31.

Hillson, D. A. (1998). Project risk management: future developments. *Int. J. Project Business Risk Mgt.* 2(2):181–195.

Hillson, D. A. (September 1998). Risk management for the new millennium. *Project Manager Today* 10(9):8–10.

Hillson, D. A. (June/July 1999). Business uncertainty: threat or opportunity? *ETHOS* (13):14–17.

Hillson, D. A. (1999). Developing effective risk responses. In: *Proceedings of the 30th Annual Project Management Institute Seminars, Symposium*, presented in Philadelphia, USA, 11–13 October 1999.

Hillson, D. A. (February 1999). Managing risk—the critical factor in successful project management. *Cost Engineer* 37(1):11–12.

Hillson, D. A. (September 1999). Project risk management—where now? *Risk and Continuity* 2(3):21–24.

Hillson, D. A. (Summer 1999). Risk and faith: contradictory or complementary? *Faith in Business Q.* 3(2):8–12.

Hillson, D. A. (2000). Benchmarking risk management capability. In: *Proceedings of the 3rd European Project Management Conference*, presented in Jerusalem, Israel, 12–14 June 2000.

Hillson, D. A. (2000). Project risk management—where next? In: *The Project Management Yearbook 2000*. High Wycombe, Bucks, UK: Association for Project Management (APM), pp. 55–59.

Hillson, D. A. (September 2000). Project risk—identifying causes, risk and effects: *PM Network* 14 (9):48–51.

Hillson, D. A. (2001). Effective strategies for exploiting opportunities. In: *Proceedings of the 32nd Annual Project Management Institute Seminars, Symposium* (PMI 2001), presented in Nashville, USA, 5–7 November 2001.

Hillson, D. A. (2002). *Defining Professionalism: Introducing the Risk Management Professionalism Manifesto*. High Wycombe, UK: PMProfessional Solutions.

Hillson, D. A. (July–December 2002). Critical success factors for effective risk management. Four-part series. *PM Rev*.

Hillson, D. A. (2002). Extending the risk process to manage opportunities. *Int. J. Project Management* 20(3):235–240. [Also in *Proceedings of the 4th European Project Management Conference* (PMI Europe 2001), presented in London, UK, 6–7 June 2001.]

Hillson, D. A. (2002). Extending the risk process to manage opportunities. *Projects, Profits* 2(11):19–24.

Hillson, D. A. (2002). The Risk Breakdown Structure (RBS) as an aid to effective risk management. In: *Proceedings of the 5th European Project Management Conference* (PMI Europe 2002), presented in Cannes, France, 19–20 June 2002.

Hillson, D. A. (2002). Using the Risk Breakdown Structure (RBS) to understand risks. In: *Proceedings of the 33rd Annual Project Management Institute Seminars, Symposium* (PMI 2002), presented in San Antonio, USA, 7–8 October 2002.

Hillson, D. A. (2002). What is risk? Towards a common definition. *InfoRM, J. UK Inst. Risk Management* (April):11–12.

Hillson, D. A. (March 2003). A little risk is a good thing. *Project Manager Today* 15 (3):23.

HM Government Cabinet Office Strategy Unit. (2002). Risk: improving government's capability to handle risk and uncertainty. Report ref. 254205/1102/ D16. Crown Copyright, London.

HM Treasury. (1993). Managing risk, contingency for defence works projects. HM Treasury CUP Guidance Note No. 41.

HM Treasury. (June 1994). Risk management guidance note.

HM Treasury. (June 1999). Procurement Guidance No. 6. Financial aspects of projects [supersedes CUP Guidance Notes Nos 15, 25, 38, and 41].

Hulett, D. T., Hillson, D. A., Kohl, R. (February 2002). Defining risk: a debate. *Cutter IT J*. 15 (2):4–10.

Ingebretson, M. (2002). In no uncertain terms. *PM Network* 16 (12):28–32.

Institute of Chartered Accountants in England and Wales (ICAEW). (1998). *Financial Reporting of Risk: Proposals for a Statement of Business Risk.* London: ICAEW.

Institute of Chartered Accountants in England and Wales (ICAEW). (1999). *Internal Control: Guidance for Directors on the Combined Code.* London: ICAEW.

Institute of Chartered Accountants in England and Wales (ICAEW). (1999). *No Surprises: The Case for Better Risk Reporting.* London: ICAEW.

Institution of Civil Engineers (ICE) and Faculty, Institute of Actuaries. (1997). *Risk Analysis, Management for Projects (RAMP).* London: Thomas Telford.

Jafaari, A. (2001). Management of risks, uncertainties and opportunities on projects: time for a fundamental shift. *Int. J. Project Management* 19(2): 89–101.

Jobling, P. E. (1997). Lessons for risk management from major projects. *Int. J. Project Business Risk Management* 1(3):255–270.

Jones, E. F. (2000). Risk management—why? *PM Network* 14(2):39–42.

Jones, M. E., Sutherland, G. (1999). *Implementing Turnbull: A Boardroom Briefing.* London: Institute of Chartered Accounts in England and Wales (ICAEW).

Kähkönen, K., Artto, K. A., eds. (1997). *Managing Risks in Projects.* Bury St. Edmonds, UK: E & FN Spon.

Kelly, P. (2002). Risk decisions under uncertainty. *InfoRM, J. UK Inst. Risk Management* (October):12–14.

Kliem, R. L., Ludin, I. S. (1997). *Reducing Projects Risk.* Aldershot, UK: Gower.

Kratz, L., Thomason, A. (1999). *Strategic Investment Decisions—Harnessing Opportunities, Managing Risks.* London: Financial Times/Prentice Hall.

Kwak, Y. H. (2001). Risk management in international development projects. In: *Proceedings of the 32nd Project Management Institute Annual Seminars, Symposium* (PMI 2001), presented in Nashville, USA, 1–10 November 2001.

Leach, L. P. (2001). Putting quality in project risk management. Part 1: Understanding variation. *PM Network* 15(2):53–56.

Leach, L. P. (2001). Putting quality in project risk management. Part 2: Dealing with variation. *PM Network* 15(3):47–52.

Lewis, A. (1999). Ten tips for successful risk management. *Project Manager Today* 11(10):38–40.

MacCrimmon, K. R., Wehrung, D. A. (1988). *Taking risks: The management of uncertainty.* New York: Macmillan Press.

Martin, P., Tate, K. (1998). Team-based risk assessment. *PM Network* 12(2):35–38.

Martin, P., Tate, K. (2001). *A Step by Step Approach to Risk Assessment.* Cincinnati, OH: Martin-Tate Training Associates LLC.

McCrae, M., Balthazor, L. (2000). Integrating risk management into corporate governance: the Turnbull guidance. *Risk Management: Internat. J.* 2(3):35–45.

Miller, R., Lessard, D. (2001). Understanding and managing risks in large engineering projets. *Int. J. Project Management* 19(8):437–443.

Murray, K. (1998). Risk management: beyond the textbooks. *PM Network* 12(6):53–57.

Newland, K. E. (1995). Benefits of risk analysis and management. *Project* 5(5):7–10.

Newland, K. E. (1997). Benefits of project risk management to an organisation. *Int. J. Project Business Risk Mgt.* 1(1):1–14.

O'Reilly, P. (1998). *Hamessing the Unicorn—How to Create Opportunity and Manage Risk*. Aldershot, UK: Gower.

Pender, S. (2001). Managing incomplete knowledge: why risk management is not sufficient. *Int. J. Project Management* 19(2):79–87.

Pritchard, C. (2001). *Risk Management: Concepts and Guidance*. Arlington, VA: ESI International.

Project Management Institute. (2000). *A Guide to the Project Management Body of Knowledge (PMBoK®)*. 2000 ed. Philadelphia: Project Management Institute.

Raftery, J. (1994). *Risk Analysis in Project Management*. Bury St. Edmonds, UK: E & FN Spon.

Renn, O. (1998). Three decades of risk research: accomplishments and new challenges. *J. Risk Res.* 1(1):49–71.

Ritchie, R., Marshall, D. (1993). *Business Risk Management*. London: Chapman, Hall.

Rosa, E. A. (1998). Metatheoretical foundations for post-normal risk. *J. Risk Res.* 1(1):15–44.

Rowe, W. R. (1988). *Anatomy of Risk*. Malabar, FL: Kreiger Publishing.

Royer, P. (2000). Risk management: the undiscovered dimension of project management. *Project Management J.* 31(1):6–13.

Royer, P. (2002). *Project Risk Management: A Proactive Approach*. Vienna, VA: Management Concepts Inc.

Sadgrove, K. (1996). *The Complete Guide to Business Risk Management*. Aldershot, UK: Gower.

Schuyler, J. (2001). *Risk and Decision Analysis in Projects*. 2d ed. Philadelphia: Project Management Institute.

Shackleton, J. (1997). Business risk management. *Technical Focus*, Issue 10. London: Institute of Chartered Accountants in England and Wales (ICAEW).

Simon, P. W., Hillson, D. A., Newland, K. E., eds. (1997). *Projects Risk Analysis, Management (PRAM) Guide*. High Wycombe, Bucks, UK: APM Group.

Simons, R. (May–June 1999). How risky is your business? *Harvard Business Rev.* 85–94.

Smith, P. G. (2003). A portrait of risk. *PM Network* 17(4):44–48.

Stewart, R. W., Fortune, J. (1995). Application of systems thinking to the identification, avoidance and prevention of risk. *Int. J. Project Management* 13(5):279–286.

Stock, M., Copnell, T., Wicks, C. (1999). *The Combined Code: A Practical Guide.* London Gee Publishing.

Teale, D. (2001). Successful Project Risk Assessment in a Week. London: Hodder, Stoughton.

Thompson, P. A., Perry, J. G., eds. (1992). *Engineering Construction Risks.* London: Thomas Telford.

UK Association for Project Management. (2000). *Project Management Body of Knowledge.* 4th ed. High Wycombe, Bucks, UK: APM.

UK Office of Government Commerce (OGC). (2002). *Management of Risk— Guidance for Practitioners.* London: Stationary Office.

US Department of Defense. (1989). *Risk Management—Concepts and Guidance.* Fort Belvoir, VA: Defense Systems Management College.

US Department of Defense. (2000). *Risk Management Guide for DoD Acquisition.* 3rd ed. Defense Acquisition University, Defense Systems Management College. Fort Belvoir, VA: DSMC Press.

Vose, D. (2000). *Risk Analysis—A Quantitative Guide.* 2d ed. Chichester, UK: John Wiley.

Ward, S. C., Chapman, C. B. (2003). Transforming project risk management into project uncertainty management. *Int. J. Project Management* 21(2):97–105.

Watkins, M. D., Bazerman, M. H. (March 2003). Predictable surprises: the disasters you should have seen coming. *Harvard Business Rev.* 81(3):72–80.

Watson, R., Ramsay, C. (2001). Focusing project risk management to enhance business performance. *Project Manager Today* 13(4):25–28.

Wideman, R. M., ed. (1992). *Project and Program Risk Management: A Guide to Managing Project Risks and Opportunities.* Newtown Square, PA: Project Management Institute.

Index

T - #0174 - 101024 - C0 - 229/152/18 [20] - CB - 9780824748081 - Gloss Lamination